Drug Design
Cutting Edge Approaches

Drug Design
Cutting Edge Approaches

Edited by

Darren R. Flower
The Edward Jenner Institute for Vaccine Research, Newbury, UK

RS•C
ROYAL SOCIETY OF CHEMISTRY

The Proceedings of the meeting on Cutting Edge Approaches to Drug Design held on 13 March 2001 at the Scientific Societies Lecture Theatre, London, UK

Special Publication No. 279

ISBN 0-85404-816-2

A catalogue record for this book is available from the British Library

Published by The Royal Society of Chemistry,
Thomas Graham House, Science Park, Milton Road,
Cambridge CB4 0WF, UK

Registered Charity Number 207890

For further information see our web site at www.rsc.org

Typeset in Great Britain by Vision Typesetting, Manchester
Printed and bound by Athenaeum Press Ltd, Gateshead, Tyne & Wear

Preface

The application of computational sciences to pharmaceutical research is a discipline whose time has come. A tranche of techniques, both old and new, have recently matured into potent weapons in the war against disease. Molecular informatics – computational chemistry or molecular modelling, bioinformatics, and cheminformatics – has reached new heights of sophistication and utilitarian value within drug discovery. As an initiative to further foster and disseminate understanding of molecular informatics within the wider pre-clinical research environment, the organising committees of the Biological and Medicinal Chemistry Sector (BMCS) and the Molecular Modelling Group (MMG) of the Industrial Affairs Division of the Royal Society of Chemistry (RSC) inaugurated a series of one day meetings to address the subject. Highly technical, highly specific meetings that cover certain methodological aspects of the discipline are quite common, but we felt need for a broader and more accessible kind of conference that would serve as a gentle introduction to cutting edge approaches to drug design. This book is the proceedings of our first such meeting.

The pharmaceutical industry is a hugely profitable global business: the total annual worldwide sales for all human therapeutic drugs is about $350 billion, while the farm livestock health market is worth about $18 billion and the annual sales for the companion animal health market is approximately $3 billion. To put these huge numbers into context: $350 billion is comparable to the yearly gross national product of Taiwan, the Netherlands, or Los Angeles County. Drug sales are increasing at about 5% a year, while the vaccine market, currently worth a modest $5 billion a year, is increasing at about 12% per annum. The annual global investment in R&D is around $30 billion, up from $2 billion in 1980. As a proportion of sales, average R&D expenditure has risen from 11.4% in 1970 to 18.5% in 2001. The merged GlaxoSmithKline is, at least in terms of market capitalisation, now amongst the top few largest companies in the world, yet controls less than 10% of the pharmaceutical market.

The structure of the global pharmaceutical market is highly biased. Over 50% of all marketed drugs target G-Protein coupled receptors. This includes a quarter of the 100 top-selling drugs, which generate sales of over $16 billion per year. Many of the top one hundred GPCR targeted drugs are so-called block-busters each earning over $1 billion dollars a year. The biggest sellers have, however, been anti-ulcer drugs that have dominated the market place for most of the last 25 years. SmithKline Beecham's Tagamet, launched in 1977, was followed by Glaxo's Zantac (launched 1983), followed by Astra's proton pump inhibitor Losec, whose global sales peaked at $6.2 billion. Putting aside these blockbusters the 'average' drug struggles to recoup its development costs. Indeed, two out of three marketed drugs fail to yield a positive return on investment.

After a long relaxed period of sustained profitability, the industry now faces a drive towards increased efficiency. The emphasis is now firmly on shortening time-to-market, yet tightening by regulatory bodies has increased the time it takes to approve each new NCE: 19 months in 2001 up from only 13.5 months in 1998. Estimates of the attrition rate within pharmaceutical R&D varies widely: figures quoted lie somewhere between 0.25 and 0.001 depending how one does the calculation. Only about 1 in 12 compounds in development reaches the market. Competition has also increased dramatically and so has the concomitant rate of mergers and acquisitions. Should current trends continue, within 5 to 10 years five companies will control about 80% of the pharmaceutical market.

Yet 40% of human disease remains incurable and many existing therapies are far from ideal. The nature of illness has, at least in the West, changed out of all recognition over the last century, and can be expected to do so again during the next hundred years. Thus the challenge to modern medicine, of which the pharmaceutical industry is a key component, has never been greater, yet neither has the technology available to address it. The post-genomic revolution – genomics, transcriptomics, and proteomics – compounded by High Throughput Screening, and the coming revolution of lab-on-a-chip super-synthesis, will deliver an unprecedented information explosion. It is only through informatic strategies that we will be able to manage and fully exploit this data overload.

The first Cutting Edge Approaches to drug design was held on March 12 2001. The meeting opened with a barn-storming performance by one of the big beasts of structure based drug design: Professor Sir Tom Blundell. This was followed by a talk by Dr Jon Terrett of Oxford Glycosystems, deputising for Dr Andy Lyall, OGS's Head of Informatics. Dr Darren Green, of GlaxoSmithKline, spoke next on the subject of Virtual Screening, followed by Dr Dave Brown of Pfizer, who described the role of X-ray crystallography in drug design. In the afternoon, we had a talk by Dr Iain McLay from GlaxoSmithKline on lead optimisation methods followed by Dr Andy Davis talking about the resurgent role of Physical Organic Chemistry in drug discovery. The day was finished off by three talks detailing applications of informatic strategies: Dr Pascal Furet (Novartis) discussed Kinase inhibitors, and Dr Peter Hunt (Merck, Sharp & Dohme) & Dr Frank Blaney (GlaxoSmithKline) discussed drug design problems in G-protein coupled receptor research.

Before we came to put these proceedings together, Dr Furet declined to contribute. Later, it became apparent that, that for various reasons, Dr Terrett,

Dr Brown and Dr Blaney would also be unable to contribute to the writing of this book. In order to compensate for this, I prevailed on Professor Teresa K Attwood, incipient grand dame of British bioinformatics, to help me describe the importance of integrated bioinformatics within G-protein coupled receptor research target discovery. My own group contributed a review of an exciting development in drug discovery research: the application of computational methods to the design of vaccines. I have also included a introductory chapter, which, apart from plumbing the depths of my own ignorance, attempts to put the other chapters into some kind of context, while trying to introduce some of the concepts that will be explained later in more detail. In writing these proceedings, we have tried to stay close to the ideal of the original meeting by attempting to balance technical accuracy with accessibility and readability for the non-specialist. Readers can judge for themselves if we succeeded.

Thanks are, of course, due to all the speakers, and their co-authors, for their astounding and outstanding efforts. I should also like to extend my thanks to the other organisers of the meeting: Dr Nicola Aston (GlaxoSmithKline, Chair), Dr Terry Hart (Novartis), both representing the BMCS, and Dr Chris Snell (Novartis). Of course, the meeting itself could not have happened without the organisational brilliance of Elaine Wellingham, to whom ineluctable thanks are due. I should also like to thank Alan Cubitt, Janet Freshwater, and the rest of the staff of RSC books, without whose help this excellent tome would never have seen the light of day.

As we have said, Cutting Edge Approaches to Drug Design (CEAtoDD) was the first of an on-going series of one-day lectures. We have already held CEAtoDD II and are planning CEAtoDD III, which will be held on March 2003. For up to date information, please visit the web-site for these meetings (currently at URL: http://www.jenner.ac.uk/CEAtoDD/CEAtDD.htm). Alternatively, visit the Molecular Modelling Group web page (URL: http://www.rsc.org/lap/rsccom/dab/ind006.htm).

Dr Darren R Flower
The Edward Jenner Institute for Vaccine Research

Contents

Molecular Informatics: Sharpening Drug Design's Cutting Edge

Darren R. Flower

EDWARD JENNER INSTITUTE FOR VACCINE RESEARCH,
COMPTON, BERKSHIRE RG20 7NN, UK

1 Introduction

The word 'drug', which derives from the Middle English word '*drogge*', first appears in the English language during the 14th century and it has, at least during the last century, become, arguably, one of the most used, and misused, of words, becoming tainted by connotations of misuse and abuse. The dictionary definition of a drug is: 'a substance used medicinally or in the preparation of a medicine. A substance described by an official formulary or pharmacopoeia. A substance used in the diagnosis, treatment, mitigation, cure, or other prevention of disease. A non-food substance used to affect bodily function or structure.' Even within the pharmaceutical industry, possessed, as it is, by a great concentration of intellectual focus, the word has come, in a discipline-dependent way, to mean different things to different people. To a chemist a drug is a substance with a defined molecular structure and attributed activity in a biological screen or set of screens. To a pharmacologist a drug is primarily an agent of action, within a biological system, but typically without a structural identity. To a patent lawyer it is an object of litigious disputation. To a marketing manager it is foremost a way to make money. To a patient – the pharmaceutical industry's ultimate end-user – a drug is possibly the difference between life and death.

Unmet medical need is, then, a constant stimulus to the discovery of new medicines, be they small molecule drugs, therapeutic antibodies, or vaccines. This unmet need has many diverse sources, including both life-threatening conditions – such as arise from infectious, genetic, or autoimmune disease – and other conditions that impinge deleteriously upon quality of life. The division between the causes of disease is seldom clear cut. Genetic diseases, for example, can be roughly divided between those resulting from Mendelian and multifactorial inheritance. In a Mendelian condition, changes in the observed phenotype arise from mutations in a single dominant copy of a gene or in both recessive copies. Multifactorial inheritance arises from mutations in many different genes,

1

often with a significant environmental contribution. The search for genes causing Mendelian disorders has often been spectacularly successful. Multifactorial diseases, on the other hand, have rarely yielded identifiable susceptibility genes. The identification of NOD2 as causative component for Crohn's disease[1] has been hailed as a major technical breakthrough, leading, or so it is hoped, to a flood of susceptibility genes for multifactorial diseases. Unfortunately, the mode of inheritance in many multifactorial diseases is probably so complex that the subtle interplay of genes, modifier genes, and causative multiple mutations, which may be required for an altered phenotype to be observed, will, for some time yet, defy straightforward deduction.

Heart disease, diabetes, and asthma are all good examples of multifactorial disroders. Asthma, in particular, is, arguably, one of the best exemplars of the complex influence of environmental factors on personal wellbeing. It is a major health care problem affecting all ages, although it is not clear if the disease is a single clinical entity or a grouping of separate clinical syndromes. Asthma is a type I, or atopic, allergic disease, as contrasted with type II (cytotoxic), type III (complex immune), or type IV (delayed type). The word 'asthma', like the word 'drug', first appears in English during the 14th century. It derives from the Middle English word *asma*: a Medieval borrowing from Latin and Greek originals, although the incidence of allergic disease has been known since ancient times.[2,3] It is a condition marked by paroxysmal or laboured breathing accompanied by wheezing, by constriction of the chest, and attacks of gasping or coughing. It is generally agreed, that, over the past half-century, the prevalence of asthma, and type I allergies in general, particularly in western countries, has increased significantly. The reasons for this are complex, and not yet fully understood. Clearly, improvements in detection will have made a significant contribution to the increased apparent incidence of asthma, and other allergies, as is seen in many other kinds of condition, although this will only make a partial contribution to the overall increase. Other causative factors include genetic susceptibility; increased allergen exposure and environmental pollution; underlying disease; decreased stimulation of the immune system (the so-called hygiene or jungle hypothesis); and complex psycho-social influences. This final class includes a rich and interesting mix of diverse suggested causes, including the increasing age of first time parents, decreased family size, increased psychological stress, the increase in smoking amongst young women, decreases in the activity of the young, and changes in house design. The last of these, which includes increased use of secondary or double glazing, central heating, and fitted carpets has led to a concomitant increase in the population of house dust mites such as *Dermatophagoides farinae* and *Dermatophagoides pteronyssinus*, which are believed to be key sources of indoor inhaled aero-allergens.

Amongst the rich, developed countries of the first world – the pharmaceutical industry's principal target population – some of the most pressing medical needs are, or would seem to be, a consequential by-product of our increasingly technologized, increasingly urbanized personal lifestyles. These include diseases of addiction or over-consumption, those that characterize the West's ageing population, and those contingent upon subtle changes in our physical environment.

Certain diseases have increased in prevalence, while the major killers of preceding centuries – infectious diseases – have greatly diminished in the face of antibiotics, mass vaccination strategies, and improvements in hygiene and public health. In 1900, the primary causes of human mortality were influenza, enteritis, diarrhoea, and pneumonia, accounting between them for over 30% of deaths. Together, cancer and heart disease were responsible for only 12% of deaths. Today, the picture is radically different, with infectious disease accounting for a nugatory fraction of total mortality. Chronic diseases – the so-called 'civilization diseases' – account, by contrast, for over 60% of all deaths.

Many of these diseases, and indeed many other diseases *per se*, are preventable, and the development of long-term prophylactics, which may be taken over decades by otherwise healthy individuals, is a major avenue for future pharmaceutical exploration. Hand in hand with the newly emergent discipline of pharmacogenetics, the development of prophylactics offers many exciting opportunities for the active prevention of future disease. As Benjamin Franklin inscribed in *Poor Richard* in 1735: 'An ounce of prevention is worth a pound of cure'. However, for drugs of this type, problems common in extant drugs will be greatly magnified. 'Show me a drug without side effects and you are showing me a placebo,' a former chair of the UK's committee on drug safety once commented. As pharmaceutical products, of which Viagra is the clearest example, are treated more and more as part of a patient's lifestyle, the importance of side effects is likely to grow. A recent study concluded that over 2 million Americans become seriously ill every year, and over 100,000 actually die, because of adverse reactions to prescribed medications. A serious side effect in an ill patient is one thing, but one in a healthy person is potentially catastrophic in an increasingly competitive market place. If the industry is able to convince large sections of the population that it has products capable of preventing or significantly delaying the onset of disease, then financially, at least, the potential market is huge. Whether such persuasion is possible, and who would bear the cost of this endeavour, only time will tell.

Important amongst civilization diseases are examples that arise from addiction and over-consumption. While obesity undoubtedly has a genetic component, it also results from a social phenomenon, with a significant voluntary component, related in part to improvements in the quality and availability of food. Likewise, diseases relating to the addiction to drugs of misuse (tobacco, alcohol, and other illegal drugs, such as heroin or cocaine) give rise to both direct effects – the addiction itself – and dependent pathological impairment, such as lung cancer or heart disease. There is a need to intervene both to address and to mitigate the behaviour itself, primarily through direct drug treatment, with or without psychological counselling, and to address its resulting harmful physiological by-products. Caring for these consequent phenomena has now becoming a major burden on health services worldwide. As individuals, people find dieting difficult and giving-up strongly addictive substances, such as tobacco, even more difficult; pharmaceutical companies are now beginning to invest heavily in the development of anti-obesity drugs and nicotine patches, *inter alia*, as an aid to this endeavour. For example, the appetite supressant anti-obesity drug Reductil or Sibutramine – a serotonin, norepinephrine, a dopamine reuptake inhibitor – has recently been licensed by the National Institute for Clinical Excellence in the UK. Vaccines are also being developed to alter the behavioural effects of addictive drugs such as nicotine and cocaine.[4,5] Xenova's therapeutic vaccine TA-NIC, a treatment for nicotine addiction, has recently entered Phase I clinical trials to test the safety, tolerability and immunogenicity of the vaccine in both smokers and non-smokers. TA-NIC is thought to be the first anti-nicotine addiction vaccine to be clinically tested. Other therapies for nicotine addiction include skin patch nicotine replacement, nicotine inhalers or chewing gum, or treatment with the nicotine-free drug Bupropion. A Xenova anti-cocaine addiction vaccine, TA-CD, is currently in Phase II clinical development. We shall see, as time passes, that this type of direct pharmaceutical intervention, targeting the process of addiction rather than just treating its outcome, will doubtlessly increase in prevalence.

Box 1 *A Global Plague*

From its original introduction into Europe at the close of the 15th century, partly as a treatment for disease, the success of tobacco as a recreational drug has been astounding. Today, smoking can be justly called a global plague. It is the number one cause of respiratory disease and the single most preventable cause of death in the industrialised west. Some estimates indicate that worldwide smoking leads to more deaths per annum than AIDS, alcohol, car accidents, homicide, and suicide. Current figures would suggest that approximately 1 in 6 people in the world smoke: about 1.1 billion smokers out of a total of 6.0 billion. Of these, 50% will die prematurely from tobacco-related illness. Half will die in middle age with an average loss of life expectancy of 20–25 years. This means that in excess of 500 million, or about 10% of the existing population, will die from smoking related diseases: 27% from lung cancer, 24% from heart disease, 23% from chronic lung diseases, such as emphysema. The remaining 26% will die from other diseases including other circulatory disease (18%) and diverse other cancers (8%). Although its incidence amongst men has slowly decreased since the late 1980s, lung cancer remains the most prevalent cause of cancer deaths in the USA causing approximately 85% of bronchogenic carcinoma. It remains a deadly disease with 5 yr survival rates of only 14%. Approximately 17 million smokers in the USA alone attempt to quit each year.

In the First World, approximately one third of all people aged fifteen years and up smoke, with the percentage increasing sharply in Asia, Eastern Europe and the former Soviet States. Consumption trends indicate that smoking prevalence is reducing in developed countries (down 1.5% per annum in the United States, for example) while increasing in less developed countries (up 1.7% per annum). Based on current trends, the World Health Organisation estimates the death toll from smoking will rise to 10 million people per year by 2025. Currently two million deaths occur each year in developed countries and 1 million deaths occur each year in less developed countries. By 2025, this ratio will alter to 3 million deaths per year in developed countries and 7 million deaths per year in less developed countries. In 1950, 80% of the men and 40% of the women in Britain smoked, and tobacco deaths were increasing rapidly. There have been 6 million deaths from tobacco in Britain over the past 50 years, of which 3 million were deaths in middle age (35–69). There are still 10

Continued on p. 5

million smokers in Britain, of which about 5 million will be killed by tobacco if they don't stop. World-wide, there were about 100 million tobacco deaths in the 20th century, but if current smoking patterns continue there will be about 1 billion in the 21st century. The harmful effects of smoking have been well understood since at least the middle of the 19th century[6–10] but it was only with the solid epidemiological evidence of Richard Doll in 1950 that the link between smoking and lung cancer firmly and finally established. In the following fifty years, the links between smoking and innumerable other diseases have become clear.

In passing, we might mention that so-called diseases of over-consumption are only recalling some of the environmental disease effects prevalent in earlier ages. Heavy smoking has similar effects on the lungs to the conditions experienced by people living in countries in the cold northern climes during earlier eras. For example, dwellers in Iron Age roundhouses, Anglo Saxon and early Medieval great halls lived in large communal environments, within these domestic settings, and contended continually with large open fires, creating a high particulate atmosphere. The physiological effects of such exposure would recreate those of a heavy smoker. Yet, in other some respects their health was surprisingly good, their diet compensating, at least in part, for other factors. For example, meat – in the form of beef, mutton, and pork – was the principal component of the Anglo Saxon diet. Meat, obtained from lean, free range animals, contained, in those times, three times as much protein as saturated, and thus cholesterol bearing, fat; a ratio reversed in modern factory farmed animals. Height is often taken as an indicator of the efficacy of diet, and the Anglo Saxons were, unlike, say, the diminutive Georgians or Victorians, as tall, at least as a population, as people at the beginning of the 21st century.

The ageing population apparent in western countries is, amongst other causes, a by-product of the increased physical safety of our evermore comfortable, urbanized, post-industrial environment. Together with decades of enhanced nutrition and the effects of direct medical advancement in both medicines and treatment regimes, this has allowed many more people to exploit their individual genetic predisposition to long life. Estimates based on extant demographic changes would suggest that by 2050 the number of the super-old, *i.e.* those living in excess of 100 years, would, within the USA, be well in excess of 100,000. In terms of its implications for drug discovery, this has led to a refocusing of the attention of pharmaceutical companies onto gerantopharmacology and the diseases of old age. Examples of these include hitherto rare, or poorly understood, neurodegenerative diseases, such as Parkinson's disease, or those conditions acting *via* protein misfolding mechanisms, which proportionally affect the old more, such as Alzheimer's disease. The prevalence of stroke is also increasing: approximately 60,000 people die as the result of a stroke annually in England and Wales and approximately 100,000 suffer a non-fatal first stroke. However, the relative proportion of young people suffering a stroke has also increased. Here, 'young' refers to anyone under 65, but stroke is not unknown in people very much younger, including infants and children. Indeed, 250 children a year suffer a stroke in the United Kingdom. This disquieting phenomenon may, in the era of routine MRI scans, simply reflect the greater ease of successful detection amongst the young as well as the old.

Looking more globally – though the danger is still real in developed western countries – new or re-emergent infectious diseases, such as AIDS or tuberculosis, pose a growing threat, not least from those microbes exhibiting drug and antibiotic resistance. As the world appears to warm, with weather patterns altering and growing more unpredictable, the geographical spread of many tropical infectious diseases is also changing, expanding to include many areas previously too temperate to sustain these diseases. The threat from infectious disease, which we have seen has been largely absent for the last 50 years, is poised to return, bringing with it the need to develop powerful new approaches to the process of anti-microbial drug discovery.

From the foregoing discussion, we can identify a large array of new, or returning, causes of human disease, which combine to generate many accelerating and diversifying causes of medical need. These come from infectious disease, which have evolved, with or without help from human society, to exhibit pathogenicity, but also from diseases of our own creation, such as those resulting directly, or indirectly, from addiction or substance abuse, to other disease conditions, which have not previously been recognized, or have not been sufficiently prevalent, due to our ageing population or changing economic demographics. Patterns of disease have changed over the past hundred years and will change again in the next hundred. Some of these changes will be predictable, others not. Medical need is ever changing and is always at least one step ahead of us. Thus the challenge to medicine, and particularly the pharmaceutical industry, has never been greater, yet neither has the array of advanced technology available to confront this challenge. Part of this is experimental: genomics, proteomics, high throughput screening (HTS), *etc.*, and part is based on informatics: molecular modelling, bioinformatics, cheminformatics, and knowledge management.

2 Finding the Drugs. Finding the Targets

Within the pharmaceutical industry, the discovery of novel marketable drugs is the ultimate fountainhead of sustainable profitability. The discovery of candidate drugs has typically begun with initial lead compounds and then progresses through a process of optimization familiar from many decades of medicinal chemistry. But before a new drug can be developed, one needs to find the targets of drug action, be that a cell-surface receptor, enzyme, binding protein, or other kind of protein or nucleic acid. This is the province of bioinformatics.

2.1 Bioinformatics

Bioinformatics, as a word if not as a discipline, has been around for about a decade, and as a word it tends to mean very different things in different contexts. A simple, straightforward definition for the discipline is not readily forthcoming. It seeks to develop computer databases and algorithms for the purpose of speeding up, simplifying, and generally enhancing research in molecular biology, but within this the type and nature of different bioinformatic activity varies widely. Operating at the level of protein and nucleic acid primary sequences, bioinformatics is a branch of information science handling medical, genomic and biological information for support of both clinical and more basic research. It deals with the similarity between macromolecular sequences, allowing for the identification of genes descended from a common ancestor, which share a corresponding structural and functional propinquity.

Box 2 *What is Bioinformatics?*

Bioinformatics is one of the great early success stories of the incipient informatics revolution sweeping through the physical sciences. Bioinformaticians find themselves highly employable: indeed many eminent computational biologists have had to re-badge themselves with this particular epithet. Their services are much in demand by biologists of most, but not yet all, flavours. But what is bioinformatics? One definition is 'Conceptualizing biology in terms of molecules (in the sense of physical chemistry) and then applying 'informatics' techniques (derived from disciplines such as applied mathematics, computer science, and statistics) to understand and organize the information associated with these molecules, on a large scale'. A more tractable definition than this, which seems more uninterpretable than all embracing, is 'the application of informatics methods to biological molecules'. Many other areas of computational biology would like to come under the bioinformatics umbrella and thus get ready access to grant funding, but the discipline is still mostly focused on the analysis of molecular sequence and structure data.

Bioinformatics, as do most areas of science, relies on many other disciplines, both as a source of techniques and as a source of data (see Figure 1). Bioinformatics also forms synergistic links with other areas of biology, most notably genomics, as both vendor and consumer. In the high throughput post-genomic era, bioinformatics feeds upon these data rich disciplines but also provides vital services for data interpretation and management, allowing biologists to come to terms with this deluge rather than being swamped by it. It is still true that bioinformatics is, by and large, concerned with data handling: the annotation of databases of macromolecular sequences and structures, for example, or the classification of sequences or structures into coherent groups. Prediction, as well as analysis, is also important, not least in trying to address two of the key challenges of the discipline: the prediction of function from sequence and the prediction of structure from sequence (see Figure 2). Although these two are intimately linked, there is nonetheless still an important conceptual difference between them. One can discern three main areas within the traditional core of bioinformatics: one dealing with nucleic acid sequences, one with protein sequences, and one with macromolecular structures (see Figure 3).

At the very heart of bioinformatics is the Multiple Sequence Alignment (see Figure 4). With it, one can do so much: predict 3D structure, either through homology modelling or *via de novo* structure; identify functionally important residues; undertake phylogenetic analysis; and identify important motifs and thus develop discriminators for the membership of protein families. The definition of a protein family, the key step in annotating macromolecular sequences, proceeds through an iterative process of searching sequence, structure, and motif databases to generate a sequence

Continued on p. 8

corpus, which represents the whole set of sequences within the family (see Figure 5). Motif databases, of which there are many, contain distilled descriptions of protein families that can be used to classify other sequences in an automated fashion. There are many ways to characterize motifs: through human inspection of sequence patterns, by using software to extract motifs from a multiple alignment, or using a program like MEME to generate motifs directly from a set of unaligned sequences. A motif or, more likely, a set of motifs defining the family can then be deposited in one of the many primary motif databases, such as PRINTS, or secondary, or derived, motif database, such as INTERPRO (see Figure 6). This brief digression into the nature of bioinformatics has been very much a simplification, as is readily seen in Figure 7, which shows some of the greater complexity that is apparent in a less drug design-orientated view of the discipline.

Within the drug discovery arena bioinformatics equates to the discovery of novel drug targets from genomic and proteomic information. Part of this comes from gene finding: the relatively straightforward searching, at least conceptually if not always practically, of sequence databases for homologous sequences with, hopefully, similar functions and roles in disease states. Another, and increasingly important, role of bioinformatics is managing the information generated by micro-array experiments and proteomics, and drawing from it data on the gene products implicated in disease states. The key role of bioinformatics is, then, to transform large, if not vast, reservoirs of information into useful, and useable, information.

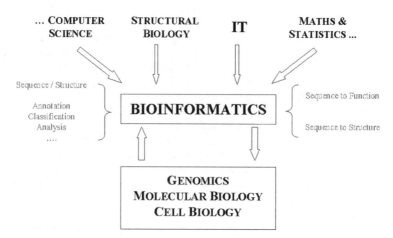

Figure 1 *Bioinformatics in its Place. Core bioinformatics makes a series of synergistic interactions with both a set of client disciplines (computer science, structural chemistry, etc.) and with customer disciplines, such as genomics, molecular biology, and cell biology. Bioinformatics is concerned with activities such as the annotation of biological data (genome sequences for example), classification of sequences and structures into meaningful groups, etc. and seeks to solve two main challenges: the prediction of function from sequence and the prediction of structure from sequence*

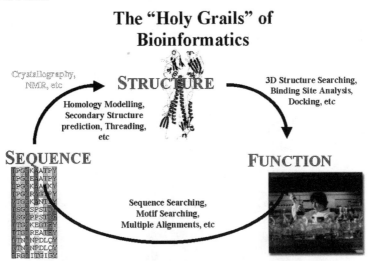

The "Holy Grails" of Bioinformatics

Figure 2 *The 'Holy Grails' of Bioinformatics. Core bioinformatics seeks to solve two main challenges: the Holy Grails of the discipline. They are the prediction of Structure from Sequence, which may be attempted using secondary structure prediction, threading, or comparative modelling, and the prediction of Function from Sequence, which can be performed using global homology searches, motif databases searches, and the formation of multiple sequence alignments. It is also assumed that knowing a sequence's structure enables prediction of function. In reality, all methods for prediction of function rely on the identification of some form of similarity between sequences or between structures. When this is very high then some useful data is forthcoming, but as this similarity drops a conclusion one might draw becomes increasingly uncertain and even misleading*

Academic bioinformaticians sometimes seem to lose sight of their place as an intermediate taking, interpreting, and ultimately returning data from one experimental scientist to another. There is a need for bioinformatics to keep in close touch with wet lab biologists, servicing and supporting their needs, either directly or indirectly, rather than becoming obsessed with their own recondite or self referential concerns. Moreover, it is important to realize, and reflect upon, our own shortcomings. Central to the quest to achieve automated gene elucidation and characterization are pivotal concepts regarding the manifestation of protein function and the nature of sequence–structure and sequence–function relations. The use of computers to model these concepts is limited by our currently limited understanding, in a physico-chemical rather than phenomenological sense, of even simple biological processes. Understanding and accepting what cannot be done informs our appreciation of what can be done. In the absence of such an understanding, it is easy to be misled, as specious arguments are used to promulgate over-enthusiastic notions of what particular methods can achieve. The road ahead must be paved with caution and pragmatism, tempered, as ever, by the rigour for which the discipline is justly famous.

One of the most important recent trends has been the identification of so-called '*druggable*' receptors. As databases of nucleic acid and protein sequences

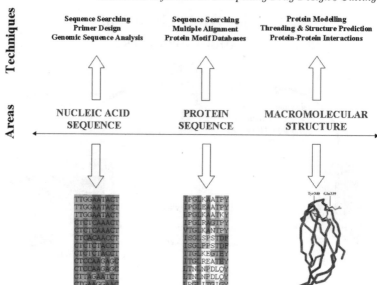

Figure 3 *Subdisciplines within Bioinformatics. Within bioinformatics there are several sub-disciplines: nucleic acid and protein sequences, and macromolecular structures. Each has its own unique techniques*

and structures have become available on an unprecedented, pan-genomic scale, attention has turned to the ability of these databases to properly facilitate the comparison of active sites across a huge variety of related proteins, and thus allow the informatician to better select and validate biological targets, better control drug selectivity in the design process, and better verify binding hypotheses in a highly parallel mode. What is, then, a druggable target? This is dependent on the nature of drug one is interested in, as clearly a short acting, injectible drug is somewhat different to a long acting, orally bioavailable pharmaceutical agent. The average G-Protein coupled receptor (GPCR), with its small, hydrophobic, internal binding site and important physiological role, is an archetypal druggable receptor. Tumour necrosis factor receptor is, on the other hand, not such a target, despite its important role in the body, as it contains no easily discernible drug-binding site. That is not to say that useful drugs can not be designed that act against it, interfering with its biological activity by blocking its action on other proteins, but it is not, in itself, obviously druggable. So by druggable, we tend to mean, to a first approximation, proteins exhibiting a hydrophobic binding site of defined proportions, leading to the development of drugs of the right size and physicochemical properties. The term druggable, then, relates, in part, to the structure of the receptor, and has another component related to the provenance of a protein family as a source of successful drug targets; that is to say, how useful have similar, related proteins been, historically, as drug targets. Estimates of the number of druggable receptors vary, as have estimates of the number of genes in the human genome. As this review is being written, estimates of gene number are converging away from the initial post-genomic 30–40,000 to

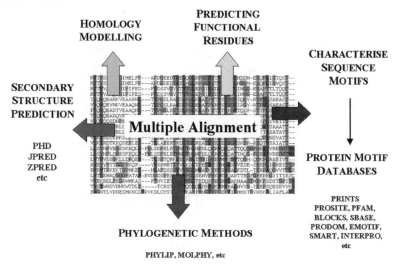

HOMOLOGY MODELLING

PREDICTING FUNCTIONAL RESIDUES

CHARACTERISE SEQUENCE MOTIFS

SECONDARY STRUCTURE PREDICTION

Multiple Alignment

PHD JPRED ZPRED etc

PROTEIN MOTIF DATABASES

PRINTS PROSITE, PFAM, BLOCKS, SBASE, PRODOM, EMOTIF, SMART, INTERPRO, etc

PHYLOGENETIC METHODS

PHYLIP, MOLPHY, etc

Figure 4 *Multiple Sequence Alignments: the Heart of Core Bioinformatics. The multiple sequence alignment lies at the heart of the bioinformatic discipline. It enables a wide range of disciplines and the accuracy of many techniques, such as those mentioned here, is heavily dependent on the accuracy of multiple sequence alignments. Examples include comparative modelling, the prediction of functional residues, secondary structure prediction (whose success is greatly enhanced by multiple sequence data), phylogenetic methods, and the deduction of characteristic sequence motifs and motif, or protein family, databases, of which many are listed here*

a more realistic 65–70,000. My own view is that this may well still prove an underestimate. Nonetheless, this puts the number of druggable receptors somewhere in the region of 2000 to 4000. Of these, about 10% have been extensively examined to date, leaving many, many receptors left to explore. Beyond the human genome, there are other druggable receptors now receiving the attention of pharmaceutical companies. Bacteria, fungi, viruses, and parasites are all viable targets for drug intervention. As the number of antibiotic resistant pathogens increases, the hunt for new antimicrobial compounds, and thus the number of druggable microbial receptors, will also expand.

2.2 Cheminformatics

Cheminformatics, named somewhat awkwardly by comparison with bioinformatics, is a newly emergent discipline that combines the decades old discipline of chemical information management, which includes substructure searching for example, with areas from molecular modelling, such as QSAR. It also subsumes other areas, such as the informatic aspects of compound control, amongst others. To a certain extent, it seeks to mirror the knowledge management aspects of bioinformatics and deals with the similarities and differences between chemical compounds rather than between protein or nucleic acid sequences.

One of the driving forces for the growth of cheminformatics has been the need

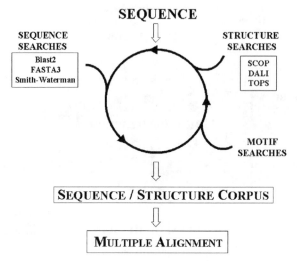

SEQUENCE

Figure 5 *Construction of a Sequence/Structure Corpus. The building of a multiple sequence alignment begins with the identification of a sequence/structure corpus. In an ideal case, this should contain all related sequences and structures related to the seed sequence of interest. The process is iterative and brings together the results of three types of searches: global sequence searches (Blast, FastA, or a parallel version of Smith-Waterman); searches against motif databases such as InterPro or PRINTS; and searches for similar 3-D structures using full model searches, such as DALI, or topology searches, such as TOPS. Once a search has converged and no more reliable sequences can be added, then the final corpus has been found and a multiple alignment can be constructed*

to support the design and analysis of high throughput screening efforts. While chemical information was, historically, concerned primarily with the cataloguing of chemical compounds in private or public collections, cheminformatics is an intellectually more active area, bringing research methods to bear on the subject. It can be assumed, with some certainty, that the greater the number and the greater the range of compounds to test the greater the likelihood of successfully identifying lead compounds. This is, of course, only true if the compounds we test are, in some meaningful sense, different from one another.

Currently, there is no generally agreed quantitative definition of chemical similarity. Many proposed methods exist, each with different strengths and weaknesses. However, choice of an appropriate measure of similarity is important: while no single method is necessarily much better than other good methods, methods which are clearly inappropriate do exist. One of the most widely used is based on mapping fragments within a molecule to bits in a binary string. It has been shown that bit strings provide a non-intuitive encoding of molecular size, shape, and global similarity; and that the behaviour of searches, based on bit-string encoding, has a significant component that is non-specific.[11] This leads one to question whether bit string based similarity methods have all the features desirable in a quantitative measure of chemical similarity.

Many molecular similarity measures have been suggested, including cal-

MOTIF SEARCHING

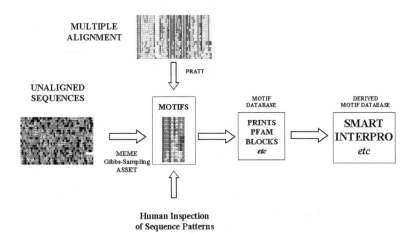

Figure 6 *Motif Databases. Motif, or protein family, databases are one of the most fruitful and exciting areas of research in bioinformatics. An underlying assumption of such databases is that a protein family can be identified by one or more characteristic sequence motifs. Such motifs can be identified in three ways. One, by direct human visual inspection of one or more protein sequences. Two, using unaligned sequences as input to programs such as MEME, which can automatically perceive statistically significant motifs. Or, thirdly, from aligned sequence using a motif identification approach such as PRATT. The resulting set of one or more motifs becomes the input into a motif, or pattern, database, of which PRINTS is an example. A derived database, such as SMART or InterPRO, can then be built on top of one or more individual motif databases*

culated properties based on representations of molecular structure at both the 2-dimensional level (topological indices and constitutional descriptors) and 3-dimensional level (properties derived from MO calculations, surface area and volume, or CoMFA).[12] Other descriptors include measured physical properties or biological activities. The number of potential quantities available is daunting. Of these many alternatives, which are the most appropriate descriptors? We might wish to choose as descriptors those properties that we feel we understand. A familiar quantity such as the octanol/water partition coefficient – Log*P* – might be a better choice, say, than a particularly obscure and poorly character-ized topological index. In the context of drug discovery, we might wish to concentrate on those descriptors that afford us some mechanistic insights into the basis of biological activity at a particular receptor. Nonetheless, similarity remains a difficult concept. On what basis can comparisons be made? There is no obvious criterion by which one can determine if one selection of compounds is better than any other. There is, ultimately, no 'gold standard' by which to judge the performance of different similarity measures. There is no consensus between chemists, or computer algorithms, and there isn't one between receptors either. There is no universally applicable definition of chemical diversity, only local, context-dependent ones. The only correct set of rules would be those that a

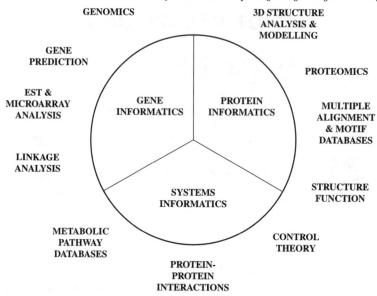

Figure 7 *The Place of Bioinformatics in Pharmaceutical Research. Within the pharmaceutical industry, and related areas, bioinformatics can be sub-divided into several complementary areas: gene informatics, protein informatics, and system informatics. Gene informatics, with links to genomics and MicroArray analysis, is concerned, inter alia, with managing information on genes and genomes and the in silico prediction of gene structure. Protein informatics concerns itself with managing information on protein sequences and has obvious links with proteomics and structure–function relationships. Part of its remit includes the modelling of 3-D structure and the construction of multiple alignments. The third component concerns itself with the higher order interactions rather than simple sequences and includes the elaboration of functional protein–protein interactions, metabolic pathways, and control theory*

receptor chooses to select molecules: but these will vary greatly between different receptors. This has not discouraged the development of a large literature – comparing methods, primarily in the context of justifying the apparent superiority of a method that the authors have developed; these are often large, complex, yet discombobulatingly terse papers which assaults the reader with the weight of information rather than the arguments of sweet reason. In the final analysis, we are in the realm of relative values; the success and failure of different measures is largely dependent on the context in which they are used, without any particular one consistently out-performing the others.

An area, within the domain of cheminformatics, where these issues are less problematic is 3D-database searching and the identification of pharmacophores. The pharmacophore is an important and unifying concept in drug design which embodies the notion that molecules are active at a particular receptor because they possess a number of key features (*i.e.* functional groups) that interact favourably with this receptor and which possess a geometry complementary to it. To a first approximation, this is a deterministic approach: a molecule either will or will not fit a pharmacophore. Like many words used in science, as in life generally, 'pharmacophore' has many meanings. Some use it to describe somewhat vague models of the environment within a ligand binding site. It is more useful, however, to give it a more specific meaning: an ensemble of interactive functional groups with a defined geometry. Although the medicinal chemistry literature is littered with pharmacophore definitions there are few general compendia.

It is possible to derive pharmacophores in several ways: by analogy to a natural substrate or known ligand, by inference from a series of dissimilar biologically active molecules (the so-called Active Analogue approach), or by direct analysis of the structure of a target protein. Most pharmacophores tend to be fairly simple two, three, or four point (*i.e.* functional group) pharmacophores, although some incorporate more elaborate features such as best planes and regions of excluded volume. Over-specifying a pharmacophoric pattern through the use of restrictive substructure criteria will limit the overall diversity and novelty of hits. In an ideal pharmacophore, the generality of functional groups does not restrict structural classes while the pharmacophore geometry supplies discriminating power to the search.

Having derived a pharmacophore model there are, in general, two ways to identify molecules that share its features and may thus elicit a desired biological response. First, *de novo* design, which seeks to link the disjoint parts of the pharmacophore together with fragments in order to generate hypothetical structures that are chemically reasonable yet typically wholly novel. The second is '3D database searching', where large databases comprising three-dimensional structures are searched for those that match to a pharmacophoric pattern. The principal advantage of this approach over *de novo* design is the ability to identify extant molecules that can be obtained ready-made or synthesized by a validated method. This is clearly rather more efficient and economic than attempting the synthesis of speculative novel molecules. It approximates to a guided screening process whereby a set of molecules is identified for biological testing which are believed to be good candidates for activity. This contrasts with true random screening where no assumptions are made about structures to be tested and potentially large numbers of compounds are screened indiscriminately. This may be appropriate in the case of high throughput screens, but in circumstances where these are not available it is necessary to prioritize compounds to be tested. In such cases, we might wish to design focused libraries or complement this by selecting from extant compounds using pharmacophore methods. Moreover, with the application of results from graph theory the celerity of 3D database searching has allowed it to establish itself as a tool in drug design with proven success in practical applications.

A single pharmacophore is unlikely to recover all compounds known to be active against a particular receptor. This is especially true for antagonists and enzyme inhibitors that bind in a number of different ways to block agonist or substrate binding. Each structurally distinct class may make its own individual subset of interactions within the total available within a binding site. Single compounds may also bind to more than one sub-site or in several different binding modes. Given the more stringent requirements of receptor activation, agonists may exhibit less diversity in binding. Thus to span the structural diversity and different binding modes exhibited by antagonists and other ligands, many pharmacophores may be required to characterize fully the structural requirements of a given receptor or pharmacological activity.

Although 3D database searching is a directed approach, there is always a need to test a reasonable number of molecules which fit a pharmacophore model. Although a particular compound may fit the pharmacophore, reflecting receptor complementarity, its activity is not guaranteed. It may possess unfavourable physical properties, overall lipophilicity for example. Likewise, it may penetrate excluded volumes within the receptor or introduce the pairing of like charges. For hits, ranges of activities obtain, including unexpected enhancements as advantageous additional interactions are made with the receptor. 3D-database searching will ideally identify compounds with properties outside that of the set of molecules used to define the pharmacophore. This allows for the identification of novel chemical structures and molecular features leading to both increased and decreased activity.

2.3 Lead Discovery

Traditionally, lead compounds have possessed key properties, such as activity at a particular receptor or enzyme, but are deficient in others, such as selectivity, metabolic stability, or their pharmacokinetic profile. New leads have arisen either as a result of serendipity (see Box 3), or from analogy to the structures of known compounds. These may be natural ligands – enzyme substrates or receptor agonists – or they may be extant pharmacological agents – inhibitors or antagonists.

Although there is no doubt that such an approach has proved successful in the past, and will continue to be successful in the future, the limitations inherent within this strategy have led many to complement it with many different alternative approaches. In the context of ethnobiology, leads may also be found by isolating active compounds from traditional herbal remedies. Many of the most successful and well-used drugs are derived in this way. Aspirin, which was originally derived from willow bark, is probably the best known example. Quinine is another good example: it is believed that during the Spanish conquest of South America it became known that Peruvian Indians were able to combat malarial fever by drinking from certain bodies of still water. It has been speculated that these bodies contained the bark of fallen Cinchona trees, a natural source of quinine. Moreover, many other drugs, including codeine, atropine, morphine, colchicine, digitoxin, tubocurarine, digoxin, and reserpine, have been discovered through the investigation of medical folklore. In comparative tests by the NCI, the hit rates for extracts from treatments derived from ethnomedicine were, in studies against cancer, about twice those of random screening. In undeveloped countries well over 80% of all treatment regimens utilize traditional cures. Assuming that they represent around 4.5 billion people, this means that something in the order of 60% of the world's population relies on

Box 3 *Serendipity in Drug Discovery*

Although the modern pharmaceutical industry spends millions of person-hours and billions of dollars to increasingly systematize the drug discovery process, many drugs, perhaps even most drugs, at least historically, have arisen through a process of serendipity. The story of Intal is particularly interesting in this context. *Ammi visnaga*, or the toothpick plant, is an Eastern Mediterranean plant, used since ancient times as an herbal remedy for the treatment of renal colic. Crystalline khellin was first isolated from the fruit of the visnaga plant in 1879, with its structure determined later in 1938. It was found to induce smooth muscle relaxation; given by mouth it causes, amongst more minor side-effects, nausea and vomiting. Yet khellin was still used clinically to dilate the bronchi in the lungs, as a coronary vasodilator to treat angina, and also had effects on asthma.

In 1955 chemists at Benger Laboratories began studying analogues of khellin as potential treatments for asthma. Compounds were screened against guinea pigs exposed to aerosols of egg white, a common animal model of asthma. At this point, asthmatic physician Roger Altounyan joined the research team. Altounyan (1922–1987) was born in Syria and later trained in medicine in England. As a child, he, and his siblings, acted as the inspiration for the Walker children in Arthur Ransome's twelve 'Swallows and Amazons' books. Seeking a more direct assay, Altounyan began to test hundreds of analogues of khellin on himself for their ability to prevent his asthma from occurring when he inhaled solubilized 'guinea pig hair and skin cells', to which he was allergic. One analogue, K84, reduced the severity of his asthmatic attack. New analogues continued to be synthesized and through tests on Altounyan, albeit at the rate of two compounds a week, an SAR was established.

In 1963 a contaminant proved highly active, and a further series of analogues was prepared. Early in 1965, Altounyan found one compound – the 670th compound synthesized during nine years work on the project – to work well for several hours. Fisons, who had merged with Benger laboratories, called it FPL670. This compound, disodium, or more commonly, sodium cromoglycate, known commercially as Intal, is used in numerous forms to treat asthma, rhinitis, eczema, and food allergy. Intal, and the follow-up drug, nedocromil sodium, both function through the stablilization of mast cells, preventing degranulation and the concomitant release of histamine and other inflammatory mediators. As a medication, Intal has strong prophylactic properties, and allows reduced dosage of steroids and bronchodilators. Clinical trials began in 1967 and Intal became the top seller for Fisons. Financially, the company was built on the success of this drug, before it eventually folded in the early 1990s, the pharmaceutical arm of the company being bought by Astra.

The PDE5 inhibitor sildenafil citrate, better known as Viagra, is the first effective oral treatment for impotence and is now prescribed in more than 90 countries worldwide. It is by far the most widely used treatment for the condition. More than 7 million men have used Viagra in the United States, with doctors there prescribing the drug over 22 million times. It is also highly effective, with up to 82% of patients experiencing benefits. As is widely known, Viagra began life in the mid–1980s as a treatment for hypertension and angina. Pfizer initially tested the drug in men without a history of coronary heart disease and then progressed to a phase II clinical trial, where it was to prove unsatisfactory as a heart medicine. At the same time, another, higher dose phase I trial showed an unexpected effect on erectile function. And the rest, as they say, is history. This is, perhaps, the finest example of happenstance working favourably within drug discovery.

traditional medicine. However, there is considerable difficulty of making scientific sense of some of the medicines. Shilajit – a brownish-black exudation found throughout Central Asia from China to Afghanistan – is a good example. This substance is a rasayana (a rejuvenator and immunomodulator), reputed to arrest ageing and prolong life, and has been used to treat ulcers, asthma, diabetes, and rheumatism. The natural history of this complex substance remains unclear despite much analytical work, but it has become apparent, after much study, that its active ingredients include dibenzo-α-pyrones, triterpenes, phenolic lipids, and fulvic acids.

A number of companies have been keen to pursue folk medicine as a source of novel drugs, often through building relationships with local healers and then testing new extracts from medicinal herbs directly in man, with resultantly high hit rates for orally active compounds. For example, Shaman pharmaceuticals have discovered a number of interesting anti-microbial compounds from tropical plants with a history of medicinal use and have collated a compendium of medicinal properties from more than 2,600 different tropical plants. After early success following this approach, the company has shifted from pharmaceutical development to nutritional supplements. Their first product to reach market is an anti-diarrhoea product that is based on a tree sap used by Amazonian healers.

Historical documents and ancient medical texts can also suggest starting points for drug discovery programmes.[13] A deep understanding of the past, as well as a grasp of present day science, is, however, required to tease out effective remedies from the endemic quackery, charlatanism, and astrological beliefs of the past. These works are typically written in either a dead language, of which Latin and Greek texts are probably the most accessible, or in an early and unfamiliar version of a modern tongue, such as medieval English. Deciphering something useful from such manuscripts requires a careful collaboration between drug discovery scientists and language experts, each possessing sufficient scholarship to understand redundant medical concepts and identify plants from inexact descriptions. Modern science has tended to regard the medicine of Galen and Aristotle, conjuring, as it does, images of blood-lettings and crude surgical techniques, as replete with errors; however, ancient herberia do contain plants of proven pharmacological effectiveness.

The scale of this should not be underestimated. Hartwell,[14] who, for example, searched hundreds of ancient texts dating back to 2800 BC, recorded over 3,000 plants used against cancer. Siliphion, a plant in the genus *Ferula*, was of significant economic importance during Hellenistic and Roman eras. As Hipprocrates records, the plant could not be cultivated, only gathered and then traded. Soranus (1st–2nd century AD), a Greek physician probably born in Ephesus, was believed to have practiced in both Alexandria and Rome and was an authority on obstetrics, gynaecology, and pediatrics. His treatise *On Midwifery and the Diseases of Women* remained influential until the 16th century. He recommended siliphion sap, administered orally, as a contraceptive; extracts of modern relatives of siliphion inhibit conception or the implantation of fertilized eggs. Pennyroyal (*Mentha pulegium* L.) is an aromatic mint plant that was used as an abortifacient in both ancient and mediaeval times. Quintus Serenus

Sammonicus, a famous collector of books (reputedly amassing 62,000 books), and the probable author of *Liber medicinalis*, wrote that a foetus could be aborted using an infusion of the plant. Latterly, the plant's active ingredient pulegone has been shown to induce abortions in animals. In this context, Ephedrine has an interesting history. It has been known in China for over 5000 years as the herbal medicine *ma huang* before its introduction into western medicine during the 1920s. The elder Pliny (23–79 AD) refers to a plant called *ephedron*, which was used to treat coughs and stop bleeding: similar to the uses that ephedrine is put to today.

Amongst the most influential of ancient medical authorities was Pedanios Dioscorides (40–90 AD), a Greek military surgeon to the Neronian armies, whose five-volume work *De Materia Medica*, written around 77 AD, was the first systematic pharmacopoeia, containing objective descriptions of approximately 600 plants and 1000 different medications. His work was much enhanced by the Islamic polymath Avicenna (980–1037), or more properly Abu Ali al-Husain ibn Abdallah ibn Sina, who wrote the monumental and epoch making *Canon of Medicine*, which was translated from Arabic into Latin and became known to the west during the 12th century. More recent texts include an important primary renaissance source, *This Booke of Soverigne Medicines* by John de Feckeneham (1515–1585), the last Abbot of Westminster was used by the Benedictine order, of which he was a member, around 1570. Renaissance folklore also included, *inter alia*, knowledge of mandrake (*Mandragora*) and deadly nightshade (*Atropa belladona*). Although this approach clearly has its own intrinsic limitations, it is able to target complicated disease states that are not conducive to high through-put models used in pharmaceutical research.

Natural products are another important source of compounds: extracts from plants, bacterial or fungal cultures, marine flora & fauna, *etc.* have all been useful in the search for novel lead compounds. Indeed, about half of the world's top 25 best-selling drugs derive from natural products. Plants in particular have proved useful sources of active pharmaceuticals. For example, there are over 120 marketed drugs with a plant origin and approximately one-quarter of all medicines prescribed in the USA, and perhaps 35% worldwide, derive from plants. Financially, this amounted to sales of over $15 billion in 1990. Paradoxically, all these highly profitable drugs have been obtained from less than 0.1% of known plant species. By contrast, compounds of marine origin form an under-exploited source of natural product drugs. Marine biology remained largely untapped as a source of compounds until the National Cancer Institute started a discovery initiative during the 1970s. This led to the discovery of a dozen or so products, including Dolastatin and Bryostatin, that are now reaching clinical trials (see Table 1). Thus the potential of natural products remains huge and, despite the considerable investment in time and resources, also remains greatly under-exploited.

By natural products, or secondary metabolites, we really mean compounds which have an explicit role in the internal metabolic economy of the organism that biosynthesized them. Several competing arguments seek to explain the existence of such seemingly redundant molecules. Of these, perhaps the most

Table 1 *Recently identified therapeutic compounds of marine origin. Compounds of marine origin that have recently started clinical trials. Compounds are sub-divided based on their organism of origin. All compounds are anti-cancer agents other than those stated explicitly below. Data from ref. 15*

Organism	Compound	Origin	Therapeutic area	Reference
Bryozoan	Bryostatin 1	Gulf of California		16
Mollusk	Kahalaide F	Hawaii		17
Sea hare	Aplyronine A	Japan		18
	Dolastatin 10	Indian Ocean		19
Sponge	Contignasterol	Papua New Guinea	Asthma	20
	Dithiocyanates	Australia	Antinematode	21
	Halichondrin B	Okinawa		22
	Hemiasterlin	Papua New Guinea		23
Tunicate	Aplidine	Mediterranean		24
	Cyclodidemniserinol trisulfate	Palau	HIV	25
	Didemnin B	Caribbean		26
	Ecteinascidin-743	Caribbean		27
	Lamellarin a 20 sulfate	Australia	HIV	28

engaging is an evolutionary one: secondary metabolites enhance the survival of their producer organisms by binding specifically to macromolecular receptors in competing organisms with a concomitant physiological action. As a consequence of this intrinsic capacity for interaction with biological receptors, made manifest in their size and complexity, natural products will be generally predisposed to form macromolecular complexes. On this basis, one might expect that natural products would possess a high hit-rate when screened and a good chance of high initial activity and selectivity. However, although potent, the very same complexity makes natural products difficult to work with synthetically. When natural products are only weak hits, they do not represent particularly attractive starting points for optimization. However, at the other extreme natural products can prove to be very potent and very selective compounds that can, with little or no modification, progress directly to clinical trials. For example, cyclosporin, FK506, and taxol have all found clinical application.

The kind of providential drug discovery described above, and especially in Box 3, requires a degree of outrageous good fortune which one can not easily factor into a business plan. Moreover, traditional drug discovery paradigms require a greater allowance of time than is currently deemed desirable. For reasons such as these, the industry has turned steadily to new methods, such as High Throughput Screening and library design, with a greater presumed intrinsic celerity. It has, at least in the past, also been contended by some that such approaches, as well as being fast, are also idiot-proof: test enough compounds quickly enough and all will be well. Surely our corporate compound banks are full of new drugs, people would say. And, if not, then in this combinatorial library or in these collections of compounds we have bought in. Time is beginning to prove this assertion wrong, though many still believe it.

In the past decade, HTS has been exploited by most major pharmaceutical companies as their principal route to the discovery of novel drugs. This technology is able to assay very large numbers of compounds in comparatively short times. At least in principle, HTS allows for the identification of novel lead compounds, in the absence of any information regarding ligand or receptor structure, for new areas of biological activity where knowledge concerning the nature of either is lacking. Notwithstanding arguments about the veracity of automated assay systems, it has generally been agreed that one can not overestimate the potential benefits of HTS technology in accelerating the drug discovery process. The degree to which this faith is justified remains unresolved. The extent to which such an approach has, or will, fulfil its potential is something that must be left to the verdict of history. Only a large, retrospective analysis of many HTS campaigns will reveal the relative success and efficiency of this approach.

To capitalize on the power of HTS it is necessary to access compounds for testing. Moreover, the number of compounds we test will profoundly influence our results. Test too few compounds and we will fail to find active compounds. Test too many and the process becomes excessively expensive and time consuming. One source of compounds is molecular libraries generated by combinatorial chemistry. Initially, designed libraries relied too heavily on overly familiar templates – benzodiazepines and peptides for example – giving rise to problems of novelty, variety, and, in the case of peptides and peptidomimetics, metabolic stability and chemical tractability. As time has progressed, these types of large, random library have given way to smaller more focused, more rational designed libraries. This is, as we shall see many times in this volume, now a mainstay of computational chemistry and cheminformatics. The other principal sources of compounds for screening are synthetic organic molecules that have accumulated in public and corporate compound banks. These are generally chemically tractable starting points, but can again suffer from lack of novelty and have unwanted additional activities.

3 Structure-based Design: Crystallography, NMR, and Homology Modelling

Dramatic and unpredictable changes in binding occur regularly as we make changes within a series of structurally related ligands. This phenomenon is well known from the results of X-ray crystallography.[29] Changes in binding mode confuse attempts to understand SAR and to design new drugs in a rational way. While we can detect changes experimentally using mutagenesis, we are often unable to anticipate them. This is where crystallography and other experimental structure determination methods, such as multi-dimensional NMR, can deliver real benefits. If one is prepared to use crystallography as people have tried to use modelling, then a experimentally based design process directly involving a fast turn-around crystallography service can deliver even greater benefits. The potential of such an approach is well known, and, indeed, many a company is based on such a paradigm.

Recently, however, interest in an ambitious meta-project – structural genomics – has grown significantly. Structural genomics is the automated and systematic analysis of the three-dimensional (3D) structures of the individual components of the proteome. For evolutionary reasons, the different proteins represented within a genome fall into discrete groups. Each group is a set of proteins related to each other at the level of their amino acid sequences. The 3D structures of each member of the set should also closely resemble each other. Thus to determine the unique 3D structures of a whole genome one needs to pick one example from each of these sequence sets. The other members of the set could then be modelled by homology techniques. The development of practical Structural Genomics progresses apace (see Box 4).

Box 4 *Structural Genomics*

We already live in a post-genomic world. The sequences of genomes from a whole variety of prokaryotic and eukaryotic organisms are now available, including, of course, the human genome. There are approximately 150 completed genome sequences, and something like 150 currently being sequenced. This is, by any means, a vast amount of information, but what is to be done with this mass of data? Clearly, the whole of biology is reorienting itself to capitalize on this *embarrasse de riches*. Well known automated and semi-automated approaches such as functional genomics, MicroArrays, and Proteomics, amongst others, are trying to decipher the biological functions of these 10,000s of genes. Structural genomics is another approach to utilizing the potential power of the genome.

For the average bacterial genome, any reasonable division would still represent hundred upon hundred of protein structure determinations, and for eukaryotic genomes, such as that of human or arabidopsis, literally thousands. So structural genomics also requires new ways to automate experimental structure determination. Traditionally X-ray crystallography has progressed through a series of stages from the very biological, or, more precisely, biochemical, through to the abstractly mathematical, visiting experimental physics on the way. First, having identified our protein of interest, we need to produce sufficient pure protein to perform the search for appropriate crystallization conditions. Once we have crystals of the protein we need to collect X-ray diffraction data from these crystals and then 'solve' the structure, which involves solving the phase problem. That is recovering the electron density within the unique part of the repeating lattice of the crystal by combining the intensities of diffracted X-rays with phase data, the other component of the Fourier transform linking real molecular electron density and the experimentally determined diffraction pattern. The final stage requires building and refining a protein model within the electron density and ultimately refining this crude model to optimize its ability to recreate the diffraction pattern. Each of these many stages represents significant obstacles to the automation of the process of protein crystallography, which has always been a highly manual undertaking. Let us look at each stage to see how the modern crystallographer is overcoming these challenging obstacles.

The production of protein is probably that aspect of structural genomics that is least specific to the automation of crystallography: these days everyone wants large quantities of pure protein. What sets it apart is a question of scale: few people want

Continued on p. 23

quite such pure protein in quite such large amounts. The development of many different high throughput protein production systems is currently underway in both academic and commercial organizations. These include both *in vitro*, or cell-free, systems and examples based on well-understood microbial systems, such as *Escherichia coli*. One significant advantage of cell-free systems is the ability to incorporate selenium containing amino acids, such as selenomethionine or the tryptophan mimic, β-selenolo[3,2-*b*]pyrrolyl-L-alanine,[30] into proteins or affect ^{15}N labelling. Selenium incorporation allows for the phasing of the protein diffraction pattern using multiwavelength anomalous diffraction, the so-called MAD technique, which offers a general approach for the elucidation of atomic structures. Likewise, the ability to label proteins with ^{15}N, or other isotopes, offers similar advantages in NMR work.

Once one has sufficient protein, the next stage in the crystallographic phase is obtaining crystals. This is one of the two main, and largely intractable, problems left in X-ray crystallography. While the other obstacle, the so-called phase problem, is slowly yielding to various different forms of attack, crystallization remains what it has always been, essentially a black art. The process of growing protein crystals, quite different from the growth of inorganic or small organic molecules, is still poorly understood and requires an empirical process of trial-and-error to determine the the relatively few idiosyncratic conditions of pH, ionic strength, precipitant and buffer concentrations, *etc.* necessary for the growth of diffraction-quality crystals. However, even this recalcitrant discipline is yielding to the power of robotics and informatics.[31] This allows many more trials to be performed and at much more accurately defined conditions than is the case for manual crystallizations. This has, in turn, led to the successful crystallization of many seemingly intractable proteins, such as several subunits from the lipocalin crustacyanin. Others have used sophisticated statistical techniques to speed the search for crystallisation conditions but cutting down the number of conditions that needed to be tested. For example, robust multivariate statistics has been used to relate variations in experimental conditions, within experimentally designed crystallization trials, to their results.[32] Although these mathematical models can not explain crystallization mechanisms, they do provide a powerful pragmatic tool allowing the setting up of crystallization trials in a more rational and more confident manner, particularly when proteins are in limited supply.

Until recently, crystal mounting has seemed that aspect of crystallography least tractable to automation. However, recent results have indicated that even this process may yield to robotics. The process of mounting a protein crystal such that it can sit comfortably in an X-ray beam is a highly interactive process requiring a prodigious feat of manual manipulation, personal dexterity, and physical adroitness. While one may learn the techniques involved it is by no means easy. However, the system developed by Muchmore *et al.* addresses most of these issues through a combination of cryogenic temperatures (which can reduce radiation damage to the crystals), intelligent software, and a high degree of robotic control.[33] Although the systems they describe have a rather Heath Robinson appearance, they are no worse than the set ups used in other high throughput regimes within the drug industry. Other problems of data collection have been solved over the last decade, with a combination of Area Detectors and high energy synchrotrons, allowing for faster collection times on smaller, more radiation fragile crystals.

Continued on p. 24

The determination of a protein structure by crystallography involves the combination of the diffraction pattern, which is obtained by allowing a focused beam of X-rays, to pass through a crystal. Each spot on the diffraction pattern (DP) represents an intensity or amplitude, and has associated with it another quantity, the so-called phase, which when combined with the intensity, through a Fourier transform, yields an electron density map. The phasing of the DP can be solved in many ways. However, unlike small molecule crystals, where phases can be determined directly from relationships between intensities, proteins require much more approximate, but nonetheless ingenious, solutions. However, in the context of structural genomics, most are undesirable. Molecular replacement requires an existing 3D model of a homologous protein, while multiple isomorphous replacement requires a trial-and-error search for heavy atom derivatives that is similar in concept, if not in scale, to crystallization trials. MAD phasing, as mentioned above, is a much better alternative, if selenium containing amino acid derivatives can be incorporated into the protein. Another approach is the development of so-called direct methods. Of these, David Sayre and colleagues,[34] have developed one of the most interesting approaches. They propose the use of ultrashort, intense X-ray pulses to record diffraction data in combination with direct phase retrieval. Their approach relies on the production of femtosecond X-ray pulses generated by free electron X-ray lasers with a brilliance $10^8 \times$ that of currently operating synchrotons. They combine these with clever manipulation of the diffraction data for single specimens to produce an accurate, phased, and interpretable electron density map.

The final part of the crystallographic process exists more or less entirely within the computer. This is the fitting and refinement process that turns the initial phase electron density into a protein structure. It is a complex computational problem involving the interaction of many different computer programs, each addressing a particular part of the refinement process: initial fitting of a protein structure to the initial phased electron density map, iterative refinement of the model, validation, *etc.* Most of these issues have been addressed singly over the last ten years and programs implementing these methods exist. The challenge then is to link them together in such a way that the process is automatic, or as near to automatic as can be, so that minimal human intervention is required.

As we have seen, many of the advances in the biochemical and biophysical stages of the crystallographic process – protein production and crystallization – will be greatly enhanced by automation, specifically the kind of robotics that have become familiar from high through screening efforts. Other technical advances will solve, or side-step, many of the inherently intractable problems of crystallography, such as the phase problem. How far it is possible to fully automate the highly interactivetechnical process of crystallography remains to be seen. Will it be success or failure? Either way it will greatly test the ingenuity of X-ray crystallographers.

The one kind of protein to which experimental techniques, such as crystallography or NMR, are not well suited are membrane proteins. This is because the maintenance of their 3D structure requires the presence of a lipid phase. The over-expression, purification, and crystallization of membrane proteins remain daunting technical obstacles, although significant progress has been made, with, for example, the recent structure determination of a GPCR.[35] The development of Structural Genomics methodology capable of providing us with membrane protein structures remains a daunting technical challenge for the future.

Structure-based design has, traditionally, followed two broad courses. One involves the solution of a novel protein structure followed by the use of computational strategies to predict or identify putative ligands that can then be assessed experimentally. We will discuss this in more detail below. The other main course is the application of a cyclical process whereby a molecule is first designed, synthesized and then crystallized as part of a protein complex. The design is performed using either computational strategies or, more empirically, using pairwise differences between compounds, as favoured by traditional medicinal chemists. This molecule is then co-crystallized, or soaked into an existing protein crystal, a difference fourier applied, and the structure refined to yield a protein–ligand complex. This process is applied iteratively until significant improvements in activity are achieved.

The other part of structure-based design is homology modelling. Here the structure of a protein is modelled using the experimentally determined structure of homologous proteins. It is now a well-established technique and automated methods that remove much of the tedium from the routine production of such models are now well known.[36,37] Problems still exist, however: the fitting together of protein domains in a multi-domain protein, the determination of the most likely conformation of protein loops, the correct positioning of amino acid side chains, flexible ligand docking – to name only a few. When one has a model, either generated by homology modelling or through an experimental technique, it becomes possible to undertake one, or more, interesting theoretical invetsigations: one can dock small or large molecules, one can design various kinds of mutant, or one can perform some kind of atomistic simulation leading, in turn, to the investigation of thermodynamic properties, principally binding (see Figure 8).

At the start of a new protein modelling project, the receptor or enzyme model will, generally speaking, be relatively poor and inaccurate. As more and more data from mutagenesis and/or ligand SAR becomes available, the corresponding model will improve in its accuracy and predictive power. In describing this process, it is useful to speak of a model's nominal resolution. Initially, the model is of low resolution: it is fuzzy and imprecise. We can visualize this as a spectrum which extends from *de novo* models (such as those generated using structure prediction methods or through the geometrically constrained modelling of membrane proteins), through sequence threading, to knowledge based homology modelling, finally reaching experimental methods (see Figure 9). In this context, it is unusual to regard structures from crystallography or NMR as models, but that is all they are after all, albeit models optimizing the ability of a 3D model to reproduce experimental measurements. As the project progresses, the model will improve as its nominal resolution increases, becoming more and more accurate and predictive. We may begin by drawing conclusions about the overall properties of bound molecules, and by the end we can make very much more specific, quantitatively accurate predictions about individual synthetic changes in ligand series.

The level of detail we draw from our analysis should match the level of detail,

PREDICTION OF
DYNAMIC PROPERTIES

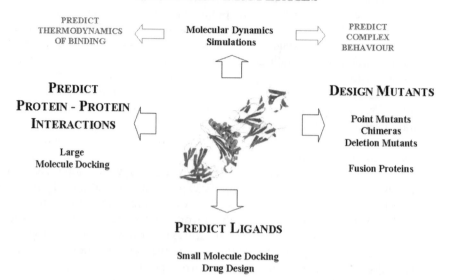

Figure 8 *Prediction of Dynamic Properties. A schematic diagram illustrating the different ways in which the synergistic interactions between macromolecular modelling, docking algorithms, and atomistic simulations can leverage solutions within the drug design process. The construction of a 3D model allows us to predict the structure of a protein molecule or complex and then to design point, chimeric, or deletion mutants or fusion proteins thereof. Applying docking algorithms to such models allows us to predict the structure of both large and small molecule complexes. The application of molecular dynamic methods to such models allows us to predict experimentally verifiable thermodynamic properties, such as the free energy of binding, of protein molecules. It also allows us to explore the 'complex' behaviour which 'emerge' from the interactions between the components in supra-molecular systems. An example of such a system might include a membrane protein embedded in the lipid phase of a membrane and interacting with solvent and dissolved small molecules*

the resolution, or fuzziness, of the model. General and qualitative at early stages – highly specific and quantitative later on. For example, when the binding mode is not well resolved, it is probably not possible to use the model directly to design ideal ligands. However, at this early stage of the development of a homology model, its gross features, which are unlikely to change, may already be clear. Analysis of binding site topography can still prove useful. We are able to make broad, yet useful, generalizations: we might be able to say, for example, that small lipophilic bases, are required for good activity. Using appropriate information, in the appropriate way and at the appropriate time, will allow us to get the most from our model. Over-interpretation at an early stage can often be misleading. At every stage, it is, ultimately, experimental validation that drives the process of model refinement.

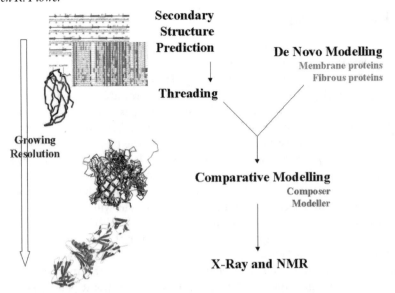

Secondary Structure Prediction

De Novo Modelling
Membrane proteins
Fibrous proteins

Threading

Growing Resolution

Comparative Modelling
Composer
Modeller

X-Ray and NMR

Figure 9 *The Nominal Resolution of Different Kinds of Protein Models. When discussing the nature of protein models, we can conveniently talk of their nominal resolution. Low resolution models are inherently fuzzy and imprecise. As we gain more and more relevant information, the model becomes more accurate and its resolution improves. It is no longer fuzzy; the sharp corners become visible. Very low resolution models correspond to predictions of a protein's secondary structure. Assuming these predictions are accurate, which is often not appropriate, then the tertiary interactions, which determine the nature of, say, the active site, remain unknown. Models produced by threading, though possessing a greater degree of tertiary verisimilitude, can be little better, offering little improvement in terms of active site structure. Likewise, the de novo modelling of membrane proteins, based, in the main, on topological constraints imposed by the 2D geometry of the membrane phase and some understanding of interacting residues, are, although for different reasons, equally imprecise. The highest resolution comes from models derived by X-ray, multidimensional NMR, and other direct biophysical techniques. Although, the prevalent view assumes, albeit subconsciously, that experimental structures are not models but reality. While this view is somewhat justified when talking in relative terms, it is not strictly correct. Both X-ray and NMR attempt to model sets of constraints, such as the difference between observed and calculated structure factors or sets of NOE distance constraints, and are prone to all manner of random and systematic errors. Intermediate between these extremes is the area of comparative protein modelling, exemplified by automated systems such as Composer[36] and Modeller.[37] This improvement in model quality, as we move up the resolution scale, gives rise to a corresponding improvement in its predictivity. We can draw conclusions about the overall properties of molecules from low resolution models and, hoepfully, make very specific, quantitative predictions about the effects of individual synthetic changes from high resolution models. The level of detail we draw from our analysis should match the level of detail, the resolution, or fuzziness, of the model. General and qualitative at early stages – highly specific and quantitative later on. Using appropriate information, in the appropriate way, and at the appropriate time, allows us to get the most from our model. Over-interpretation at an early stage can be misleading: at every stage, it is experimental validation that drives the process of model refinement*

4 Library Design

Parallel synthesis and combinatorial chemistry now allow the medicinal chemist to supplement the synthesis of individual compounds with the use of the compound library as a tool in drug discovery. These concepts can be traced back at least as far as Hanak's work in the 1960s,[38] and probably much further, implicit as they are in the very concept of synthesis itself. These techniques, together with automated screening methodologies, offer great benefits and have generated concomitant levels of interest and excitement within the pharmaceutical industry.

In principle at least, the ability to test the large numbers of compounds generated by combinatorial methods should allow for a great acceleration in the discovery of new medicines. However, unless one can achieve this both more cheaply and more quickly, then any potential benefit will lost. These issues are now seldom considered important, as more and more faith is put in the reliability and veracity of high throughput screening. However, the instability and chemical reactivity of many tested compounds can severely compromise many an HTS campaign.[12] Moreover, the intrinsic trade-off between speed and time in the high throughput equation reduces the value of each data point generated by each technique. This greatly affects the type of analysis one can reliably perform on HTS data: it has long been the desire to extract meaningful SAR from large data sets, but currently the random component within such sets precludes this.

The capacity of combinatorial chemistry to generate large numbers of compounds can either be directed towards the generation of a universal library (a large, generalized library containing innumerable diverse compounds) or towards the construction of smaller, more focused libraries of more similar structures. Design of the first of these types is well met by methods from the emergent discipline of cheminformatics. The second type requires knowledge of the structural requirements for activity at a particular kind of receptor; knowledge which can come from SAR or from an understanding of the receptor structure itself.

Lying somewhere between the extremes of a universal library and a receptor-focused library is the idea of a targeted library, which is directed against a defined class of biological targets, mostly likely a protein family such as the GCPRs or protein kinases. For a particular protein family, assuming some of its members have been well studied in the past, a large number of compounds, active at these proteins, should already exist. These will be either commercially successful compounds or, more likely, one of many other compounds that, while they are potent agonists or antagonists, never reached the market. The result is a wealth of chemical knowledge regarding the structural features inherent in this class of ligands. This has led to the use of such 'privileged fragments' to construct combinatorial libraries. The commonalities apparent in the structures of different receptors are thus reflected in the many structural features shared between the small molecule agonists or antagonists of the particular receptor family.

These libraries are consequently expected to possess a much-increased probability of yielding active ligands from screening. There is enough evidence in the open literature, and within individual pharmaceutical companies, to suggest that this can be, at least in a pragmatic sense, a quite successful strategy; though rigorous tests of the assumptions that underlie this approach are not common. A library targeted at a particular ligand series should be expected to yield more and better hits than a universal library. This type of library should, in its turn, do better than libraries targeted at other distinct types of receptor. Such a test, run over sufficient, and sufficiently different, libraries for a statistically valid number of targets, is probably unlikely ever to be realized, except as the by-product of on-going high throughput screening campaigns. The word on the street, however, is not overly favourable: the combination of large-scale combinatorial chemistry and high throughput screening is no longer viewed as the saviour of the drug discovery process. The view that 'mindlessly' constructing a large enough number of molecules and out will pop not just a lead but a development candidate has proved both naïve and shortsighted. In many ways it was replacing thought with action. This has certainly been the attitude of many a drug discovery manager, keen to overcome the strange inability of his medicinal chemists to find novel, highly active and selective drugs with a good Drug Metabolism and Pharmacokinetic (DMPK) profile in an afternoon's work. Their view would be to obviate the need for the intellectual input of skilled chemical knowledge by generating sufficient numbers of compounds.

There is clearly still a place for large libraries, but as a tool alongside many others rather than as an all-conquering doomsday weapon. One argument raised in the defence of such thinking is a comparison with quasi-evolutionary strategies: the SELEX-powered generation of aptamers, or phage display, or, in a slightly different context, the affinity maturation of antibodies, all of which, typically, give rise to a number of highly 'potent' compounds. Not necessarily compounds that could, themselves, be used as drugs, but nonetheless affine. Here, of course, we are dealing with perhaps 10^6 times as many compounds, suggesting that if one could move to astronomically high numbers of compounds then the problems of drug discovery would disappear. Although synthetic ultra-minimization suggests such numbers are possibly achievable, problems of signal-to-noise suggest that the inherent veracity of screening would require a similar increase in sensitivity and reliability.

There has been a move to make lead-like libraries of smaller, less complex molecules, capable of exhibiting some activity in your screen or screens but with sufficient room-to-grow that they can be optimized by the addition of hydrophobicity and interactive groups while retaining acceptable ADME (Absorption Distribution Metabolism and Excretion) properties. Thus in the generation of libraries of potential leads, it is important to distinguish molecules with 'lead-like' properties from molecules from other sources.[39,40] There are also 'drug-like' leads, which may be marketed structures, such as propranolol. Natural products can also be leads. As we have said, these have high affinity but like, say, NPY or taxol, they also have high – some might say daunting – chemical complexity. Drug-like leads will have affinity greater than 0.1 μM, MW greater than 350, and a clogP greater than 3.0. Lead-like leads will often have affinity greater than 0.1 μM, MW less than 350 and a clogP less than 3.0. Natural product leads will have affinity perhaps orders of magnitude lower than 0.1 μM, MW much higher than 350, and a clogP less than 3.0.

Similar arguments related to the size and complexity of leads have led to the development of new technologies that use various highly accurate biophysical measurements, as opposed to biochemical assays, as their primary readout. 'SAR by NMR', developed by Fesik and others,[41] uses an NMR-based method by which small, low complexity organic molecules binding at proximal subsites within a protein binding site are identified, optimized, and then tethered to form highly affine ligands. Using this approach it is possible to design, say, nanomolar compounds by linking together two micromolar ligands. The needle approach of Hoffmann-la-Roche is an alternative approach to HTS that uses a similar basic concept.[42] They undertook virtual screening using the 3D structure of their target protein to search for potential low molecular weight inhibitors, which were then assayed experimentally using a carefully biased screen, with hits verified using biophysical methods. The resulting sets of inhibitors were optimized guided by the 3D structural information. Their initial set of 350,000 compounds was reduced to 3,000 molecules, which yielded 150 hits in their experimental assay. Optimization provided highly potent inhibitors, ten times better than literature precedents.

5 Virtual Screening

In the era of combinatorial chemistry and high throughput screening the analysis of virtual chemical structures has assumed a position of central importance within computational chemistry, impinging directly on the design of combinatorial libraries. Of course, all computer representations of molecules are, and always have been, virtual, but historically these representations have often corresponded to molecules that have been synthesized or ones whose synthesis has been carefully planned. Now it is possible to generate literally billions of structurally feasible molecules that exist only within large virtual libraries. How do we evaluate them? Which small subset do we keep and which larger set do we throw away?

One route, as we shall see later, is to screen them out on the basis of their size and computed physical properties. Another is to try to evaluate their potential activity either using some form of QSAR or through so-called virtual screening. This technique, as the term is most often used and understood, involves using a receptor model – a protein active site say – to evaluate molecules, *i.e.* to predict, quantitatively or qualitatively, some measure of receptor binding. There are two linked and unsolved problems that frustrate attempts to develop virtual screening methodologies: the automatic docking of ligands within the binding site and the quantitative prediction of ligand affinity. Although many methods for automated ligand docking have been suggested,[43–45] and although there have been some successful applications,[46,47] their overall performance remains relatively poor. Likewise, it remains difficult to predict reliable measures of binding affinity using protein–ligand complexes, even where experimental structures are available.[48–50] Solving these problems remains a major challenge for computational chemists.

Solutions to the docking problem must take account of the flexibility of both ligand and protein, and, if one is docking against a homology model, then one must also take account of errors in the modelled structure. This leads to a combinatorial explosion in the number of possible ways of docking an individual molecule each of which must be evaluated. To deal with this, powerful computational optimization algorithms, such as Monte Carlo or genetic algorithms, are now often employed. The work of Anderson *et al.*[51] is a recently reported attempt to bypass some of these problems. They defined a minimum set of flexible residues within the active site and thus, effectively, increasing the docking site from a single conformation to an ensemble with, it is hoped, a concomitant decrease in the bias that is inherent in the use of a single, rigid protein conformation. It is not the first, nor likely to be the last, attempt to do something of this sort.[52–54]

Likewise, the probing of the active site can have a major impact on the quality of dockings. There are two main approaches to this problem. One uses some kind of pre-generated set of favourable interaction points within the active site and tries to fit molecules to this, in a way analogous to the fitting of molecules into an initial electron density map in X-ray crystallography. There are many ways to identify these points of interaction including GRID fields[55] or Multiple Copy Simultaneous Search.[56] The alternative strategy is to evaluate a potential docking using some form of molecular mechanics energy evaluated between docked ligand and receptor. In either case one would attempt to evaluate and score, for each molecule, several different docking conformations and orientations.

As there are many ways to perform ligand docking, there are, now, many virtual screening methodologies currently in circulation, all with their own advantages and disadvantages. Most attempt to overcome the limitations of computer time by using very simple methodologies that allow each virtual small molecule structure to be docked and scored very quickly. Examples of these include GOLD[57] and DOCK,[58] amongst many more. Of course, virtual screening methods exhibit a wide range of alternative methodologies of increasing complexity, from simple rule-based scoring to what are, essentially, forms of relatively time consuming atomistic molecular dynamic simulation such as Linear Interaction Energies (LIE).[59] There has been some attempt recently to combine the results of these different approaches, of which CScore, distributed by Tripos Inc, is, perhaps, the best known. My own experience of such software would suggest that any improvement that might come from using data fusion methodologies such as this, is strongly tempered by the nature of the problem one is trying to solve. It may increase the gain of true positives in a particular screening experiement, but has much less success in producing an improved quantitative correlation with experimental data. A somewhat similar approach, which is specifically designed to produce more accurate quantitative data, has been described by So and Karplus.[60] They evaluated a variety of different methods using 30 glycogen phosphorylase inhibitors as their test set. The methods they employed covered a variety of 2D and 3D QSAR methodologies, as well as structure-based design tools such as LUDI. A jury method used to combine the different independent predictions led to a significant increase in predictivity.

The relative success of FRESNO[61,62] in the prediction of binding affinities for MHC–peptide interactions perhaps suggests that optimization of the screening function, within a chemical area or protein family, rather than the use of totally generic screening functions, may be a better route to success. Indeed, the inability to predict quantitative binding constants using simulation approaches has led many to combine calculations with some type of statistics in order to leverage model predictivity. Examples of this include COMBINE[63] and VALIDATE.[64] One of the most interesting of these approaches is PrGen.[65,66] This approach uses correlation-coupled minimization to optimize the receptor–ligand interactions for a series of ligands of known affinity so that the model becomes predictive both within, and hopefully beyond, the training set. Liaison is a program, distributed by Schrodinger Inc, which combines molecular mechanics LIE methodology with a statistical model building capacity to generate models of ligand affinity within defined ligand receptor series.

A warning note for such methods, however, comes from studies by Groom and co-workers.[67,68] By using X-ray crystallography, they determined the high-resolution crystal structures of thermolysin (TLN) soaked in high concentrations of co-solvents acetone, acetonitrile, phenol and isopropanol. Analysis of the solvent positions shows little correlation with interaction energy computed using a molecular mechanics force field or with favourable positions defined using GRID probes. However, the experimentally determined solvent positions are consistent with the structures of known protein–ligand complexes of TLN. Indeed the structure of the protein complex was essentially the same as the native apo-enzyme. This suggests that existing potential energy functions are not accurate enough to correctly model protein–ligand interactions even for the simplest ligands. Yet this approach is, nonetheless, widely used.

One of the most interesting *de novo* design methodologies to emerge within the last ten years is also an example of such a strategy. The multi-copy simultaneous search (MCSS) approach, and its derivative methods,[69,70] originally developed by Miranker,[71] uses a molecular mechanics formalism to place large numbers of small functional groups – simple ketones or hydroxyls – at favourable positions within a protein's active site. In their method, the protein sees the whole swarm of ligands but each of the functional groups only sees the protein, not each other. Dynamic Ligand Design is a truly elegant extension of this approach with powerful conceptual appeal.[72] The results of the MCSS are turned into molecules under the influence of a pseudo-potential function that joins atoms correctly accounting for stereochemistry. Their potential energy function allows atoms to sample a parameter space that includes both the Cartesian coordinates and atom type. Thus atoms can mutate into different element types and hybridizations. Subsequently, a modified version of the method was developed which used a new potential energy function, optimization by simulated annealing, and evaluation using a thermodynamic cycle.[73] An alternative *de novo* design strategy is the SPROUT suite developed in Leeds.[74] This program uses a more conventional method for transforming sets of site points, derived from pharmacophores or from inspection of an active site, into whole ligand molecules. SPROUT now has the interesting feature that, incorporating concepts from RECAP,[75] it is able to joins site points together using a set of synthetically accessible steps that links purchasable molecular fragments together using known types of reaction. This brings *de novo* design, with its notoriously synthetically inaccessible molecules, closer to the can-do world of combinatorial chemistry or parallel synthesis.

Clearly the more resources, in terms of both human and computer time, one is prepared to employ in generating and evaluating possible dockings, the more likely one is to obtain a good solution. Likewise, the more sophisticated, and thus, generally, time consuming, are our methods for evaluating the scoring phase of the virtual screening process, the more likely we are to accurately screen our virtual library (see Box 5). If we want to dock a few dozen small molecule structures, then we can afford to expend a great deal of time on this process, but if our goal is to dock a large virtual library, then the practical limitations of computing power will reduce this to a minimum.

Box 5 *Computing Resources: the Sky's the Limit*

As we have seen, one approach to various problems in computational chemistry is to simplify, as far as possible, our computing methodology. An alternative is to use high performance computing which allows for more computationally demanding approaches. For example, previous attempts to utilize molecular dynamics and other atomistic simulation methods in drug discovery have foundered on technical limitations in present day computing methods. While many methods linking thermodynamic properties to simulations are known, most of them require an unrealistically long time to evaluate free energies or other energetic quantities. For example, a basic simulation yielding a free energy of binding needs, as a minimum, about 10 nanoseconds of simulation. On the kind of simple serial machines that are generally available, this would require a computing time in the order of 300 hours per nanosecond. To simulate even small systems in a realistic manner might occupy a whole machine for several years.

To circumvent these limitations we might seek to use massively parallel implementations of molecular dynamic codes running on large supercomputers with 128, 256, or 512 processor nodes. However, the widespread use of high performance computing has often been limited by the poor availability of true supercomputers. Compac recently won the world's supercomputer crown, beating IBM's Asci White into second place. IBM had held the number one position for the last three years. The current champion is a 3,024-processor machine called Terascale, which is based at the Pittsburgh Supercomputing Centre. It can perform six trillion calculations per second, the equivalent of 10,000 desktop PCs. Such large multi-processor machines are, however, generally available only as time-shared resources, while the availability of 'home made' distributed supercomputing composed of LINUX clusters, although significant amongst bioinformaticians, has made less of an impact in chemistry.

Barabasi *et al.* have recently proposed a novel form of so called parasitic computing,[76] where one machine hijacks other target computers, through internet communication protocols, to perform components of large computational tasks. The resulting

Continued on p. 34

virtual supercomputer raises interesting ethical problems. A very similar idea is one used by Search for ExtraTerrestrial Intelligence (SETI) [http://setiathome.ssl.berkeley.edu/]. This and other 'hard' computing tasks, such as the simulation of protein folding, are now seeking solutions using so-called Peer-to-Peer computing protocols, whereby computers world wide voluntarily donate computing power. Peer-to-peer computing is defined as the sharing of resources between computers, such as processing time or spare storage space. Internet-based peer-to-peer applications position the desktop at the centre of computing, enabling all computer users to participate actively in the Internet rather than simply surfing it. Recently, computational chemists have also taken up this approach. The Cancer screensaver project [http://www.chem.ox.ac.uk/curecancer.html] is an initiative by Graham Richards and colleagues at Oxford University's Centre for Computational Drug Discovery – a 'virtual centre' funded by the National Foundation for Cancer Research. They are working with Intel and United Devices, a US distributed computing company, to perform the virtual screening of 3.5 billion compounds. The project is also aiming to use this technique to search for drugs to combat anthrax toxin.

Another alternative, although not one currently realized, is the utilization of fundamental advances in computing technologies: biological, chemical, and optical computing offers, but has yet to deliver, untold increases in computing speed. Faster still, of course, is quantum computing. Apart from the technical challenges in manufacturing such devices is the fundamentally different types of computer programming required to make them work.

Grid computing, a fundamental paradigm shift in the financial and social nature of computing, will allow all of these different approaches to the need for high performance computing to be seemlessly integrated. The term refers to an ambitious and exciting global effort needed to make this vision a reality. It will develop an environment in which individual users can access computers, databases and experimental facilities simply and transparently, without having to consider where those facilities are located. It is named by analogy with the national power transmission grid. If one wants to switch on a light or run a fridge freezer, one does not have to wait while current is downloaded first, thus Grid seeks to make available all necessary computer power at the point of need. E-Science is the first step of this process to be realized. It refers to the large-scale science that will be carried out through distributed global collaborations enabled by the Internet. Typically, a feature of such collaborative scientific enterprises is that they will require access to very large data collections, very large-scale computing resources and high performance visualisation back to the individual user scientists. The Internet gave us access to information on Web pages written anywhere in the world. A much more powerful infrastructure is needed to support e-Science. Scientists will need ready access to expensive and extensive remote facilities, to computing resources such as a teraflop computer, and to information stored in dedicated databases.

In the first phase of the Grid initiative in the United Kingdom, six projects were funded by the EPSRC: RealityGrid, Structure-Property Mapping, Distributed Aircraft Maintenance Environment, Grid-enabled Optimization and Design Search for Engineering, Discovery Net: An E-Science test-bed for high throughput informatics, and Mygrid: Directly Supporting the E-Scientist. Of these, one of the closest to our present topic is RealityGrid. The project aims to grid-enable the realistic modelling and simulation of complex condensed matter structures at the meso and nanoscale

Continued on p. 35

levels and is a collaboration between teams of physical and biological scientists, software engineers, and computer scientists. The long-term ambition of the project is to provide generic technology for grid-based scientific, medical and commercial activities. RealityGrid proposes to extend the concept of a Virtual Reality centre across the grid and links it to massive computational resources at high performance computing centres. Using grid technology to closely couple high throughput experimentation and visualization, RealityGrid will move the current bottleneck out of the hardware and back to the human mind. A twin-track approach will be employed within RealityGrid: a 'fast track' will use currently available grid middleware to construct a working grid, while a 'deep track' will involve computer science teams in harnessing leading-edge research to create a robust and flexible problem-solving environment in which to embed RealityGrid. To meet its objectives, it will utilize a computing environment built around the UK's most advanced computing technology and infrastructure.

6 QSAR

Quantitative Structure Activity Relationships, more usually referred to by the acronym QSAR, is a discipline conceptually distinct, if operationally complementary, to molecular modelling. It is a long-standing scientific area stretching back to the work of Overton.[77] It first grew to prominence within pharmaceutical research following the pioneering work of Corwin Hansch.[78] The fundamental objective of QSAR is to take a set of molecules, for which a biological response has been measured, and using statistical, or artificial intelligence methods, such as an artificial neural network or genetic algorithm, relate this measured activity to some description of their structure. The outcome, then, of a QSAR study is equations that relate, through statistically sound and hopefully predictive models, the activity, or more generally, the biological responses or physical properties, of a set of molecules to their molecular structure. These relationships can give mechanistic insights, but there is no requirement for this. Their ability to provide mechanistic explanations is dependent on the form of the particular molecular description. A QSAR equation based on $\log P$ or σ may, perhaps, and here some might argue this point, give rather more detail than ones based on topological indices, for example.

QSAR methods are widely applicable, and today focus as much on the prediction of intestinal absorption or blood–brain barrier crossing, as they used to do on the prediction of activity within a congeneric series. There are two areas of technical development with QSAR: the development of new, and hopefully improved, descriptions of molecular structure and the development of new statistical or artificial intelligence methods which can relate these descriptions to some measured biological or physical property. These molecular descriptions, when expressed as numerical variables, are commonly referred to as molecular descriptors. These can take many forms. One type includes physically measured values, such as $\log P$. Another type includes topological constants of varying complexity. These are 2D QSAR descriptors, as their calculation does not involve the 3D structures of the molecules involved. 3D QSAR has come to refer to methods, such as CoMFA and CoMSIA, which calculate a grid of interaction energies around the aligned sets of molecular structures and use those as its descriptors. The need to align molecules has proved problematic and other 3D QSAR methods, such as Almond,[79] have sought to dispense with this requirement. Descriptors abound in the literature. There are so many that it would require a book of their own just to review and compare them. On a somewhat smaller scale there is also an explosion in the number of different statistical and quasi-statistical techniques used to create QSARs, which began with multiple linear regression, and have progressed to more robust multivariate methods, artificial intelligence (AI) techniques, and decision trees, to name but a few. There has been some debate in the literature about the relative statistical accuracy of multivariate methods, such as Partial Least Squares, *versus* that of neural networks. Neural networks have suffered from four main problems: overfitting, chance effects, overtraining, or memorization, and interpretation. As new, more sophisticated neural network methods have been developed, and basic statistics applied to their use, the first three of these problems have been largely overcome. Interpretation, however, remains an intractable problem: not even computer scientists can easily visualize the meaning of the weighting schemes used by neural networks. Quite recently, an interesting new technique has made its appearance. Support Vector Machines, or SVMs, are another type of AI method capable of acting as a statistical engine. This is another kind of discriminant method, trying to divide up a property-hyperspace into regions favourable, or disfavourable, to activity.

One of the most important recent trends in QSAR has been the development of generic models that address the prediction of very generalized molecular properties. These may be physicochemical or biological in character, but rather than addressing questions of receptor potency, they address the prediction of more general transport properties such as blood–brain barrier crossing or permeability in caco-2 screens. The work of Michael Abraham and co-workers is typical of efforts in this area:[80] their approach is to generate multiple linear free energy relationships using simple descriptors, which include the solute excess molar refraction, the solute polarizability, the solute hydrogen-bond acidity and basicity, and the solute volume. As these values can be calculated directly from the structure of molecules, it becomes possible to predict barrier-crossing properties in an automated fashion.

7 Integrating the Process: Drugness

Some of the many ideas discussed above converge in the concept of 'Drugness'. This is the small molecule analogue of the druggable receptor concept that we encountered in preceding discussions. Some estimates put the number of available drug-like compounds at around 10,000. But what determines Drugness? That is to say what properties do we require in our candidate drugs? This depends, of course, on the context. For example, the properties of orally bioavailable, long acting compounds are very different from short acting, injected compounds. By Drugness, then we tend to mean the set of desirable, or drug-like, properties we would like our developing drug to possess. These are multifarious, heavily dependent on the type of drug we are aiming to create, and the stage we have reached in the drug discovery process. Nonetheless, there has been the realization that multiple molecular properties, both in terms of biological activity and physical properties, compatible with drug-like properties, can be built into the design of molecules in both a specific and a general way. This is one of the key methodological advances in the recent development of molecular modelling and particularly cheminformatics. This can work at the level of single molecules designed late in the drug discovery cycle or much earlier at the level of large combinatorial libraries and, of course, anywhere in between. We can use these concepts as a guiding principal in the selection, or purchase, of new compounds. For some, the concept of Drugness has become an all-conquering mantra, obliterating all counter-arguments. Whatever one might think about this view, Drugness nonetheless remains of crucial importance in modern-day drug discovery.

At one extreme, Drugness can be thought of as identifying those structural properties that would preclude the selection of particular compounds. We can summarize our exclusion criteria as the Good, the Bad, and the Ugly. The Good refers to retaining compounds with some desirable feature, such as the presence of certain interactive atoms or groups or an appropriate balance between acyclic and cyclic structures. For example, less than 5% of oral drugs are totally acyclic, while essentially no compound contains only cyclic bonds.

The Bad refers to potentially reactive functional groups, such as protein-reactive, bond-forming electrophiles. These compounds continue to appear in the medicinal chemistry literature, in patents, and even as clinical candidates. Such compounds include the reactive ketones and Michael acceptors that are advanced as potential magic bullets or suicide inhibitors. These inherently reactive compounds often appear as false positives in early stage drug discovery because of their covalent action. Though they can become drugs – at least under certain conditions – they remain strongly problematic compounds within structured drug discovery.

By the Ugly we mean the presence of certain features, such as certain functional groups or high complexity, which render a compound an unattractive starting point for optimization. Few, if any, databases are totally clean and lose no structures at this stage. The proportion varies, as does the type of structure screened out. For example, databases biased towards chemical reagents should, by their very nature, suffer considerably higher rejection rates than databases whose compounds have supposedly been pre-selected for their suitability for screening.

Such selection criteria, from this last category, are often referred to colloquially as Lipinski analysis:[81] the use of upper and/or lower bounds on quantities such as molecular weight (MW) or log$P$ to help tailor the *in vivo* properties of drugs. The rule of 5 developed by Lipinski predicts that good cell permeation or intestinal absorption is more probable when there are less than 5 H-bond donors, 10 H-bond acceptors, MW is less than 500, and the calculated log$P$ is lower than 5. A more careful experimental analysis of orally available, marketed drugs indicates slight differences to the Lipinski criteria, albeit for a set of small, relatively old drugs, but this analysis certainly confirms similar overall property patterns. The properties of agrochemicals – pesticides and herbicides – can be very different. Bioavailability arises in plants through a combination of potency, stability, and mobility, which is generally characterised as passive transport. For agrochemicals, MW should fall between 200 and 500, clog$P$ should be less than 4, and the number of hydrogen bonding groups should be less than 3. These criteria do not greatly diverge from Lipinski's rules of 5, but a significant difference is the requirement for an acidic pK_a. Human drugs, at least orally bioavailable ones, are biased in their properties towards lipophilic basic amines, where as acidic compounds, with their increased non-specific binding by Human Serum Albumin (HSA) and other plasma proteins, are significantly under-represented.

High throughput screening has lead to the discovery of lead compounds with higher MW, higher lipophilicity, and lowered solubility. Driven by the goal of affinity enhancement, medicinal chemistry optimization is likely to exacerbate all of these trends, as leads progress inexorably towards clinical candidates. This has led many to identify criteria for Lead-likeness, as opposed to Drugness. Leads have to meet variable, project dependent selection criteria. These may include validated biological activity in primary and secondary screens, normally against known targets, for a series of related compounds, must be patentable, and have a good initial DMPK profile. Historical analysis of leads is difficult, complicated by the bias inherent within the medicinal chemistry literature, and by the intrinsic complexity of the optimization process. Although the two chemical spaces overlap, nonetheless there appears to be real difference between lead and drug, particularly in pairwise comparisons. Property ranges for lead-like compounds can be defined: 1–5 rings, 2–15 rotatable bonds, MW less than 400, up to 8 acceptors, up to 2 donors, and a log$P$ range of 0.0 to 3.0. The average differences in comparisons between drugs and leads include 2 less rotatable bonds, MW 100 lower, and a reduction in log$P$ of 0.5 to 1.0 log units. Thus, one of the key objectives in the identification of lead-like compounds for screening, either by deriving subsets of corporate, or commercial, compound banks, or through the design of libraries is the need for smaller, less lipophilic compounds that, upon optimization, will yield compounds that still have drug-like properties.

Beyond these kinds of readily evaluated criteria, are properties, structural, as well as more broadly physicochemical, that determine the interaction of drugs with the whole organism, rather than with the binding site of their biological target, their ultimate site of action. They are usually termed DMPK, or ADME/tox, problems. These cover a range of biological functions: absorption by the gut, non-specific drug binding in the blood by human serum albumin or α-1-acid glycoprotein, and the metabolic clearance of compounds. Overall ADME/tox properties are, typically, relatively difficult to predict for very large data sets because experimental screens work *via* multiple mechanisms of active and passive transport. Nonetheless, in recent years pharmaceutical scientists working in the fields of drug absorption, pharmacokinetics, and drug metabolism have seen their responsibility grow beyond the provision of supporting data for regulatory filings of new chemical entities. Exciting advances in technology have allowed the enhanced gathering of data on absorption, distribution, metabolism and excretion, which has, in turn, allowed DMPK or ADME/tox scientists to make important contributions to the drug discovery process.

Within the development part of the overall discovery program, ADME/tox or DMPK information is, typically, crucial to registration in both early and late phases of the process. The term is often used to refer to non-clinical studies, but it is quite general, being equally applicable to pharmacokinetic and metabolic investigations in both humans and animals. A number of properties are measured for early phase animal studies. These including toxicokinetics, pharmacokinetics and absolute bioavailability in male and female examples of the toxicological species under investigation, protein binding and plasma distribution, whole body autoradiography/tissue distribution, metabolite profiles in toxicology species, pharmacodynamics, and allometric scaling. Later during drug development, both clinical and non-clinical, where the focus has shifted to human studies, such studies are intended to more fully characterize human drug disposition, particularly in the therapeutic target population. Such studies can include, *inter alia*, toxicokinetics in chronic and teratology studies, multiple dose pharmacokinetics, biliary excretion and enterohepatic recirculation, metabolite identification in toxicology species, multiple dose whole-body autoradiography/tissue distribution, and induction effects on metabolism. Measurements in clinical studies include: single/multiple dose pharmacokinetics in safety and tolerance studies, dose proportionality, mass balance and metabolite profile, pharmacokinetics in gender, age, and genetic subpopulations, drug interactions, pharmacodynamics, population pharmacokinetics and pharmacodynamics, and studies on bioequivalence.

Hitherto then, a vast amount of work was conducted on drug candidates during development, many of which had been poorly characterized in terms of ADME/tox and DMPK properties. This, at least, is clear. It is also clear that this imbalance was not optimal for the efficiency of the whole process. It is a truism, within the industry, that development is funded at a vastly higher level than discovery, and it is equally true that most compounds fail to reach the market, not for lack of potency but for reasons related to DMPK. Thus it can prove counterproductive to nominate new drugs, expend large sums, and then see a drug fail. Better, surely, to evaluate these properties, at least in part, at an early stage of discovery. Predictions and computational tools can aid in this endeavour.[82]

The key is reaching a balance between potency and pharmacokinetic properties. Lipophilicity, for example, is an important physicochemical parameter that can increase oral absorption, plasma protein binding and volume of distribution, and strongly influence processes such as pharmacokinetic properties and brain uptake, but high lipophilicity can also increase the intrinsic vulnerablity of compounds to P450 metabolism and high clearance. Increasing MW often leads to an increased potency but at the cost of poor solubility, and so on. Balancing the desired properties for gastrointestinal absorption and brain uptake is particularly difficult to address. One might think that, intrinsically, activity is the harder nut to crack, but it is no longer possible to assault it in isolation. Can we start with undruglike and become more druglike? Are we on a metastable point on the drug design hypersurface able to move towards more druglike properties or must we start, say, with good Lipinksi descriptors and improve potency from that point? On the other hand we can become too prescriptive: are we prisoners of our experience? Historical analyses of medicinal chemistry efforts indicate only what has been done, not what is possible. It is easy to make thousands, even tens of thousands, of inactive compounds. Indeed, combinatorial chemists do this with extreme regularity. We can design activity into any series and we can point to the thousands of molecules that have the correct profile for adsorption and distribution. The difficulty comes in combining the two. We must strike a balance between introducing groups that affect the correct physical properties for ADME/tox, principally lipid solubility, and for activity (see Figure 10). These are often opposed, yet the balance can also, sometimes, be beneficial, as ultra-active compounds can have non-ideal DMPK properties, and yet can still be efficacious as so little is required to act effectively at their receptor.

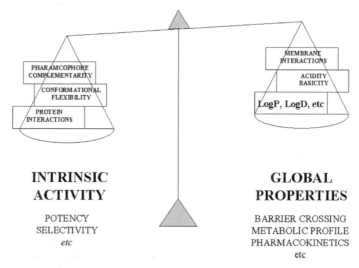

Figure 10 *Activity* vs. *Drug-like Properties: the key balancing act for drug design. The balance between intrinsic activity, and related properties such as sub-type selective, and the drug-like properties of the molecule, such as its physico-chemical profile, is the key challenge to modern-day drug design. While activity is almost entirely a product of inherent features of the drug molecule, such as flexibility, and of protein–drug interactions, albeit within the biological milieu, which imposes certain constraints that are interpretable in global terms, the drug-like properties are a mixture of both physical (barrier crossing phenomena) and protein interactions (metabolic sensitivity)*

8 Discussion

Within the ultimately profit-driven pharmaceutical industry, the discovery of novel chemical entities is the ultimate fountainhead of sustainable profitability and future commercial success. In the agrochemicals sector there is an analogous situation with regard to the identification of new herbicides and insecticides. The three informatics disciplines described above can make significant contributions to the drug discovery process, as these proceedings will amply demonstrate. Figure 11 illustrates how the three principal disciplines within molecular informatics – molecular modelling, cheminformatics, and bioinformatics – sit within pharmaceutical research. Within computational chemistry, the strong interconnectedness of its own subdisciplines is equally well promulgated by Figure 12, as are its links to other disciplines with the drug discovery process.

The crude and arbitrary way I have divided this introduction is a good illustration of how interlinked all the facets of the drug design process are, and how increasingly it is becoming dependent on the ability to process information in the large. One could have chosen alternative, equally valid, divisions of the discipline. Moreover, there are many that I have omitted: quantum mechanics and conformational searching, for example. Are those disciplines I have described truly at the cutting edge of drug design? Who can say? Must-have technologies that are the flavour of the moment can quickly lapse into obscurity pushed out by newer, sexier methods. Forgotten approaches may, as has happened often in crystallography and chemical information, be rescued from a

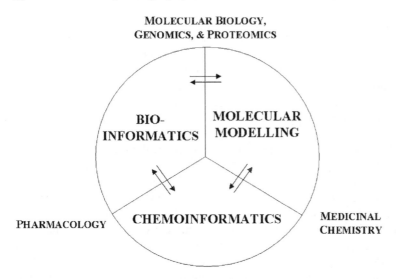

Figure 11 *The Synergy of Molecular Informatics. The three areas of molecular informatics within pharamaceutical research – molecular modelling, cheminformatics, and bioinformatics – act synergistically within drug design. Information feeds in all directions and the different areas interact differently with principal customers within the overall design and discovery process*

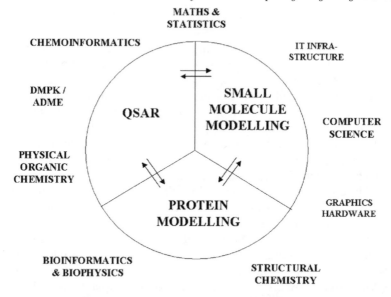

Figure 12 *The Synergy of computational chemistry. The three areas of computational chemistry – small molecule modelling, protein modelling, and QSAR – act synergistically to progress drug design. Information feeds between them and also between a set of customer and client disciplines within the pharmaceutical industry, ranging from the simple IT desktop and workstation infrastructure, through knowledge sources such as structural chemistry, to DMPK*

similar obscurity and suddenly put up in lights.

Currently, large amounts of data are being generated by a tranche of high throughput technologies: genomics, proteomics, microarray experiments, high throughput screening, *etc.* Other methods – pharmacogenomics for example – will, in time, generate an even greater volume. One of the tasks of modern drug research is to evaluate this embarrassment of riches. How much useful data is locked away? Can we reduce the set of incoherent data into useable, comprehensible information? Can we parse knowledge from this information? Can we ultimately draw out understanding from the accumulation of knowledge? One way that we can assault this problem is through computer-based molecular informatics techniques: a combination of bioinformatics, molecular modelling, cheminformatics, supplemented by knowledge management, mathematical modelling, and, as the GRID evolves, e-science. None of this is meant, of course, to replace the human part of the process. It is merely a supplement to that, albeit a powerful one: compensating for an area where the human mind is relatively weak, the fast and reliable processing of massive data sets.

People have spoken for some time now about data mining human or microbial genomes. The 'omes', of course, now abound: transcriptomes, proteomes, metabolomes, immunomes, even chemomes. We might add another, all embracing 'ome': the 'infome'. In the context of a pharmacutical company, this goes beyond the narrow confines of sequence or chemical structure data, and it is, in the broadest sense, the complement of all its biological and chemical knowledge.

It will be a great challenge to the development of knowledge management, to seek to deal with this highly homogeneous volume of information. The pharmaceutical company is probably one of the few organizations that can, within the area of molecular sciences at least, hope, through its intrinsic scale, both in terms of existing data and its willingness to invest in the future, to be able to pursue such an endeavour.

But what will we gain from this? It is worth recalling that many drugs have unexpected and unimagined lives well after their licensing and market launch. Let us take a few, high profile, examples. Viagra, originally licensed as an anti-impotence drug, can provide effective treatment for, among others, pulmonary hypertension (high blood pressure in arteries of the lungs), a disabling condition, severely limiting exercise capability, and shortening life expectancy, caused by emphysema, bronchitis, or disease in the heart valves. Thalidomide was, for a long time, a notorious drug because of its responsibility for the physical abnormalities seen in hundreds of babies born in the 1950s and 1960s. It has now been reintroduced in the battle against several diseases. The most recent has been as a medication in patients suffering myelodysplastic syndromes, conditions where the production of blood cells is severely disrupted, which often develops into acute myeloid leukaemia. In the search for drugs able to delay or reverse the terrifying effects of vCJD, two drugs, quinacrine (an anti-malarial agent) and chlorpromazine (an anti-psychotic medication) have proved promising when tested in mice. This phenomenon is called therapeutic switching. The list of similar instances where drugs have exhibited unexpected medical benefits in quite different areas to those in which they were originally licensed is very large and the scope for finding such new therapies is even larger. So large, in fact that companies, such as Arachnova Ltd (whose pipeline is based on novel patent-protected uses for existing drugs), have been formed that financially exploit the concept of therapeutic switching. There are implications inherent in this concept for the limitations of both animal models for specific diseases and for clinical trials. In the latter case, trials may be using too small a number of patients, which are too biased in their ethnicity, age, and gender to adequate cover both the effects and side effects of drug candidates, and it is an obvious challenge to the developing disciplines of pharmacovigilance and pharmacogenetics. It also provides food for thought for both pharmaceutical business strategists and regulatory bodies.

Modelling has not yet reached the limits of its usefulness or the breadth of its application. Apart from perfecting the many techniques described above, molecular modelling has many other exciting challenges facing it in the future. What, for example, are the future roles of promiscuous drugs able to act with multiple activities at several receptors simultaneously?[86] Is there a place within pharmaceutical research for compounds operating, as some believe Chinese medicines do, at sub-therapeutic values at a wide range of receptors? What role does it have to play in the development side of pharmaceutical research? Can we apply material science modelling methods to problems in formulation and drug delivery? Will molecular modelling find a role as a design tool in the emerging discipline of nanotechnology?

Box 6 *High Doses* vs. *Low Doses* vs. *No Doses*

Generally, the pharmaceutical industry is concerned with the development of compounds exhibiting high activity at a well defined set of one, or more, receptors together with a favourable profile of other relevant properties such as metabolic stability or oral bioavailability. There is typically some kind of trade-off between these properties; very high affinity can, sometimes, be sacrificed if the physical properties are so good that less compound is required to produce a therapeutic effect. However, some substances produce a biological response at an extraordinarily low concentration.[83] Known as the 'low' or 'small' dose effect, this is generally assumed to be dosed at a concentration substantially lower than the equilibrium dissociation constant of the effector–target complex. However, effects at extremely low concentrations (in the order of 10^{-19} molar) have been reported. Indeed, the phenomenon of hormesis, the occurrence of a U- or inverted U-shaped dose–response relationship, is well documented in numerous biological, toxicological, and pharmacological investigations.[84] The concept of hormesis has a long history dating back to Paracelsus (1493–1541), more properly known as Theophrastus Phillippus Aureolus Bombastus von Hohenheim, who noted in the 16th century that various toxic substances may be beneficial in small quantities. Modern research in hormesis originated over a century ago with the work of Schulz, who noted that many chemicals could stimulate growth of yeast at low doses but were inhibitory at higher levels. The concept of a general stimulation at low doses with high-dose inhibition was supported by many observations, becoming known as the Arndt-Schulz law. Despite the widespread recognition of apparent hormetic effects, the Arndt-Schulz law gradually fell into disuse for many reasons. These include high-dose toxicology testing that precludes the demonstration of low-level effects and the threat posed by hormesis to the currently accepted precautionary principle, which assumes that any dose of a chemical is potentially harmful.

The term 'hormesis' can also refer to beneficial effects from low doses of potentially harmful substances. Although there are many examples of this phenomenon from the laboratory, it remains a controversial concept and has never become widely accepted by the medical community. Many vitamins and minerals are essential for life at low doses but become toxic at higher ones. Similarly, exercise, caloric restriction, and alcohol consumption are examples of processes that are harmful in the extreme but beneficial in moderation. Likewise, a co-hormetic is a compound which at relatively 'low doses', will, in combination with some other stimulus, demonstrate increased growth while at higher 'doses' will inhibit this increased proliferation. Some view traditional Chinese medicine as acting in this way. Many Chinese take an interesting view of western *versus* eastern therapies: they will seek the help of highly potent western drugs to deal with acute disease conditions, but look to the prophylactic powers of traditional cures to keep them healthy. One explanation of this is to suggest that such complex mixtures of substances, many very active when isolated but present here in sub therapeutic concentrations, may work synergistically to effect an overall therapeutic benefit which combines mild stimulation and suppression of many, many physiological processes.

In this context, many Chinese medicines would be classed as panaceas. A 'panacea' is defined as a remedy for all diseases, evils, or difficulties; a universal medicine; a cure-all; hence, a relief or solace for affliction. In western medicine, aspirin is a good

Continued on p. 45

example of a panacea. Its uses are various: pain control, suppression of inflammation, and as an anticoagulant, thus a treatment for stroke and various cardiovascular problems. Homeopathic medicines are another treatment muted as a panacea.

The basic principle of homeopathy is 'the doctrine of similars': a remedy that causes the same or a very similar pattern in healthy subjects best treats a patient with a specific pattern of symptoms. Homoeopathic remedies are typically prepared in extremely high dilutions, where they are unlikely to contain any molecules of the original agents. Consequently, homoeopathic remedies cannot act by classical pharmacological means, and are, likewise, generally dismissed by scientific sceptics. There are many explanations of homeopathy, some strongly physico-chemical invoking complex ideas about the molecular organization of water, others evincing explanations no more believable than Star Trek technobabble. An appealing explanation is immunological in nature, invoking the concept of bystander suppression, a special kind of active inhibition, to explain the regulation induced by very low substance concentrations.

Nonetheless, it would be dangerous to totally dismiss homeopathy. In a recent meta-analysis of trial data, 18 reviews were analysed:[85] of these, six addressed the question whether homeopathy is effective across conditions and interventions. The majority of available trials reported positive results but the evidence was not wholly convincing. For oscillococcinum for influenza-like syndromes, galphimia for pollinosis, and for isopathic nosodes for allergic conditions the evidence was promising while in other areas results were equivocal.

The obvious extension of low doses is the phenomenon of the placebo effect, whereby the very act of undergoing treatment seems to aid patient recovery. Often it is taken to refer to a medicine given to please, rather than cure, a patient. Placebo, the latin for 'I shall please', is the first word of the vespers for the dead, and, as a result, during the 12th century vespers were often refered to as placebos. By the 14th century, the word had become pejorative, meaning a sycophant or flatterer, and when it entered the medical vocabulary this negative connotation remained. Most placebos are taken during double-blind clinical trials: a pharmacologically inert substance substituting for a compound under test. In such situations, patients receiving both real and imaginary drugs undergo a similar treatment regime: medical evaluation, a thorough discussion of their condition, and receive a plausible diagnosis and treatment plan. During this, patients will experience the attention, commitment, and even possibly, the respect of both nurses and medics. Perhaps the simplest explanation for this presupposes a link between a patient's psychological and physiological status: people who expect to get better are more likely to do so. One might speculate whether medical training is required in these situations. Would watching *Casualty* or *ER* accomplish something similar for an ailing patient? The healing environment, whether this manifests itself as taking a pill or interacting with some form of medically trained individual, is a powerful antidote to illness. In the words of Voltaire: 'The art of Medicine consists of keeping the patient in a good mood while nature does the healing'.

Though managers seldom realize this the most precious resource that the industry has is its staff. These are men and women of the highest educational and ethical standards, people on the whole deeply committed to the future success of their organizations. It is these people who build the future. Within this organizational structure, the molecular modeller has a special place. As with many people, their potential is limited only by those around them. The best, or most effective, kind of modeller is then, one who their colleagues, particularly medicinal chemists, will willingly listen to and be willingly influenced by. Yet influence is a two-way process, particularly within a strongly hierarchical structure as exists in many pharmaceutical companies.

It is interesting to consider how informatics impacts on the philosophy of science. There are two popular philosophical views on the nature of the scientific method. One is attributed to Karl Popper (1902–1994). This view maintains that science does not start with observations from which inductive claims are made but rather with conjectures which may subsequently be refuted by appeal to experiment but which can never be fully proven. There is congruity here with Thomas Kuhn's (1922–1996) book *The Structure of Scientific Revolutions*. Popper's interpretation supersedes the earlier inductive theory of developed by Sir Francis Bacon (1561–1626). In his assessment of modern informatics techniques Gillies[87] seeks to distinguish between competing paradigms: 'Bacon hoped that scientific theories could be generated from observations by some kind of mechanical process which places "all wits and understandings nearly on a level"'. In contrast to this view, Popper believed that the creative thinking of brilliant scientists forms scientific theories. According to Gillies, approaches to informatics fall into an arena delineated by Bacon. Indeed Bacon's emphasis on assembling experimental data into a 'table', from which inductive truths may be discovered, bears a remarkable resemblance to the use of large relational databases.

The integration of automatic synthesis and high throughput screening with informatics interpretation, in whatever form that may take, coupled to some form of automatic steering, allows any refinement to be implemented computationally rather than by purely human intervention. This frees human involvement to concentrate on the choice of search and, ultimately, the use of the information obtained. However, the idea that human creativity might be displaced conjures up considerable suspicion. Many seek to condemn informatics despite its widespread use. Some seek to demonstrate that hypothesis-free machine learning is impossible. The reality is, however, that background knowledge and hypotheses are nearly always tacitly incorporated by the choice of sample space and measurement method. The skill seems to be to integrate informatics methods properly into conventional scientific research programmes.

Much of what a modeller does can be classified as forecasting: predicting to a purpose. It is useful to draw a comparison with other kinds of forecasting, particularly weather forecasting. In the average North European country, such as Britain, which can easily experience four seasons in a day, foretelling the weather has important economic outcomes. The British Meteorological Office quotes an accuracy rate of approximately 86%. Good though this appears, it still means, on average, that one day a week the predictions are wrong. Likewise, the computational chemist can not guarantee every single prediction will be correct. If they make 50 predictions, 48 of which are correct, this is a significant achievement. However, if the most interesting molecules were part of the 2 rather than the 48, then this achievement will seem to be lessened. Weather forecasting has, like molecular modelling, seen an immense improvement on a few decades ago, when forecasting was only accurate for, say, a day ahead. Today, in no small part due to the potent combination of informatic and experimental disciplines, in this case computer simulation and satellite images, we can be confident of accurate weather forecasts up to five days in advance. In time series analysis such as this, as in molecular modelling, extreme extrapolation is rarely successful. As in all fields of science, it is important to realize the limitations of our technology, and not to seek answers to impossible questions. In this way, the benefits of informatics research can be maximized.

The attrition rate within pharmaceutical drug discovery is, from a corporate viewpoint, the main underlying problem within the process. It has been estimated that for every drug that ultimately reaches the market, about 1000 other 'projects' have been failed. Here project refers to anything from a brief biological dalliance with an assay through lead discovery projects and into a full, and highly expensive, development project. The accuracy of this 1000 to 1 ratio is open to question but this statistic neatly captures the enormous wastage rate within the industry. The perhaps 15 year long journey from exploratory biological assay development to registration and marketing is exceedingly expensive. In an attempt to overcome this, the pharmaceutical industry has over the years become temporarily fascinated by certain technologies, each promising much for the future of drug discovery, but, to some at least, each ultimately delivering little. In the 1970s and early 1980s, molecular modelling promised the Earth, or at least some early practitioners did. The technique, though useful even in those early days, was clearly overblown. During the late 1980s, the pharmaceutical industry dabbled with anti-sense as a therapeutic cure-all. In the 1990s, it has been high throughput screening, and even more recently the new informatics disciplines: chem- and bioinformatics. However, these raised expectations have been easily dashed and the jury is still out on the impact of the more recent methods. It is possible to feel cynical about the whole process. However, only by combining all relevant technologies, both biological and chemical, can one hope to move forward. No one technology, can, or could, answer all questions, solve all problems, or deliver all solutions. Only an holistic integration of all experimental and informatics skills can hope to meet the challenges we have outlined.

The pharmaceutical industry, like most organizations in the technology sector, creates the future. Old fashioned attitudes, which can still be peddled by universities, and closed-minded ways of thinking are the ultimate enemy here. The industry should engage as fully as possible with informatics disciplines: only then can true progress be made. Yet the pharmaceutical industry faces, as all industries do, financial and commercial pressures as well as scientific challenges. There is an ever-increasing need to accelerate the discovery of new drugs, bringing them to market in ever-shorter times. To do this, old-style technologies will not be sufficient. Informatics, in all its guises, will become essential to the industry's efforts to speed the discovery process.

Acknowledgements

I thank Christine Flower for expert comments on this text. I should also like to thank the following for all the information, collaboration, and discussion over the years: T K Attwood, A C T North, P C L Beverley, F Ince, P Meghani, M Christie, G Smith, D Henderson, D Elliott, S Lydford, P Shelton, G Melvin, T McInally, C McPhie, N P Gensmantel, N P Tomkinson, A Baxter, P Borrow, D P Marriott, S Thom, N Kindon, S Delaney, I Walters, I Dougall, C Law, M Robertson, L Nandi, S A Maw, S St. Gallay, R Bonnert, R Brown, P Cage, D Cheshire, E Kinchin, A M Davis, C Manners, K McKechnie, J Steele, M Robertson, P Thomas, R Austin, P Barton, G Elliott, X Lewell, S Wong, P Taylor, P Guan, K Paine, I A Doytchinova, M J Blythe, C Zygouri, H McSparron, H Kirkbride, M Hann, D V S Green, I McClay, P Willett, and F Blaney. To others I have omitted, I apologize. Without this help, the current article could not have been written.

References

1. J.P. Hugot, M. Chamaillard, H. Zouali, S. Lesag, J.P. Cezard, J. Belaiche, S. Almer, C. Tysk, C.A. O'Morain, M. Gassull, V. Binder, Y. Finkel, A. Cortot, R. Modigliani, P. Laurent-Puig, C. Gower-Rousseau, J. Macry, J.F. Colombel, M. Sahbatou and G. Thomas, Association of NOD2 leucine-rich repeat variants with susceptibility to Crohn's disease, *Nature*, 2001, **411**, 599–603.
2. E. Simons (ed.), *Ancestors of Allergy*, New York Global Medical Communications Ltd., New York, 1994.
3. J. Ring, Erstbeschreibung einer atopischen Familienanamnese im julisch-claudischen Kaiserhaus: Augustus, Claudius, Britannicus, *Hautarzt*, 1983, **36**, 470–474. [Title translates as: First report of a family anamnesis of atopy in the House of the Julio-Claudian emperors: Augustus, Claudius, Britannicus]
4. P.R. Pentel, D.H. Malin, S. Ennifar, Y. Hieda, D.E. Keyler, J.R. Lake, J.R. Milstein, L.E. Basham, R.T. Coy, J.W. Moon, R. Naso and A. Fattom, A nicotine conjugate vaccine reduces nicotine distribution to brain and attenuates its behavioral and cardiovascular effects in rats, *Pharmacol. Biochem. Behav.*, 2000, **65**, 191–198.
5. M.R. Carrera, J.A. Ashley, B. Zhou, P. Wirsching, G.F. Koob and K.D. Janda, Cocaine vaccines: antibody protection against relapse in a rat model, *Proc. Natl. Acad. Sci. U.S.A.*, 2000, **97**, 6202–6206.
6. O.S. Fowler, *Disquisition on the Evils of Using Tobacco, and the Necessity of Immediate and Entire Reformation*, Providence, S.R. Weeden, 1833.
7. W.A. Alcott, *The Use of Tobacco: Its Physical, Intellectual, and Moral Effects on the Human System*, Fowler and Wells, New York, 1836.
8. B.I. Lane, *The Mysteries of Tobacco*, Wiley and Putnam, New York, 1845.
9. J.H. Kellogg, *Tobaccoism*, The Modern Medicine Publishing Co, Battle Creek, Michigan, 1922.
10. R. Pearl, Tobacco smoking and longevity, *Science*, 1938, **87**, 216–217.
11. D.R. Flower, On the properties of bit string-based measures of chemical similarity. *J. Chem. Inf. Comp. Sci.*, 1998, **38**, 379–386.
12. D.R. Flower, DISSIM: a program for the analysis of chemical diversity, *J. Mol. Graph. Model.*, 1998, **16**, 239–253.

13. B.K. Holland (ed.), *Prospecting for Drugs in Ancient and Medieval European Texts: a Scientific Approach*, Harwood, 1996.
14. J.L. Hartwell, *Plants used against Cancer*, Quartermain publications, Lawrence Publications, Massachusetts, 1982.
15. R.C. Willis, Nature's pharma sea, *Mod. Drug Discov.*, **5** (1), 32–38.
16. P.G. Goekjian and M.R. Jirousek, Protein kinase C inhibitors as novel anticancer drugs, *Expert Opin. Investig. Drugs*, 2001, **10**, 2117–2140.
17. Y. Kan, T. Fujita, B. Sakamoto, Y. Hokama, H. Nagai and K. Kahalalide, A new cyclic depsipeptide from the Hawaiian green alga bryopsis species, *J. Nat. Prod.*, 1999, **62**, 1169–1172.
18. S. Saito and H. Karaki, A family of novel actin-inhibiting marine toxins, *Clin. Exp. Pharmacol. Physiol.*, 1996, **23**, 743–746.
19. A.B. da Rocha, R.M. Lopes and G. Schwartsmann, Natural products in anticancer therapy, *Curr. Opin. Pharmacol.*, 2001, **1**, 364–369.
20. F.R. Coulson and S.R. O'Donnell, The effects of contignasterol (IZP-94,005) on allergen-induced plasma protein exudation in the tracheobronchial airways of sensitized guinea-pigs *in vivo*, *Inflamm. Res.*, 2000, **49**, 123–127.
21. R.J. Capon, C. Skene, E.H. Liu, E. Lacey, J.H. Gill, K. Heiland and T. Friedel, The isolation and synthesis of novel nematocidal dithiocyanates from an Australian marine sponge, Oceanapia sp., *J. Org. Chem.*, 2001, **66**, 7765–7769.
22. M.H. Munro, J.W. Blunt, E.J. Dumdei, S.J. Hickford, R.E. Lill, S. Li, C.N. Battershill and A.R. Duckworth, The discovery and development of marine compounds with pharmaceutical potential, *J. Biotechnol.*, 1999, **70**, 15–25.
23. W.R. Gamble, N.A. Durso, R.W. Fuller, C.K. Westergaard, T.R. Johnson, D.L. Sackett, E. Hamel, J.H. Cardellina 2nd and M.R. Boyd, Cytotoxic and tubulin-interactive hemiasterlins from *Auletta* sp. and *Siphonochalina* spp. sponges, *Bioorg. Med. Chem.*, 1999, **7**, 1611–1615.
24. S.G. Gomez, G. Faircloth, L. Lopez-Lazaro, J. Jimeno, J.A. Bueren and B. Albella, *In vitro* hematotoxicity of Aplidine on human bone marrow and cord blood progenitor cells, *Toxicol. In Vitro*, 2001, **15**, 347–350.
25. S.S. Mitchell, D. Rhodes, F.D. Bushman and D.J. Faulkner, Cyclodidemniserinol trisulfate, a sulfated serinolipid from the Palauan ascidian *Didemnum guttatum* that inhibits HIV-1 integrase, *Org. Lett.*, 2000, **2**, 1605–1607.
26. O. Kucuk, M.L. Young, T.M. Habermann, B.C. Wolf, J. Jimeno and P.A. Cassileth, Phase II trail of didemnin B in previously treated non-Hodgkin's lymphoma, an Eastern Cooperative Oncology Group (ECOG) Study, *Am. J. Clin. Oncol.*, 2000, **23**, 273–277.
27. W.W. Li, N. Takahashi, S. Jhanwar, C. Cordon-Cardo, Y. Elisseyeff, J. Jimeno, G. Faircloth and J.R. Bertino, Sensitivity of soft tissue sarcoma cell lines to chemotherapeutic agents: identification of ecteinascidin-743 as a potent cytotoxic agent, *Clin. Cancer Res.*, 2001, **7**, 2908–2911.
28. M.V. Reddy, M.R. Rao, D. Rhodes, M.S. Hansen, K. Rubins, F.D. Bushman, Y. Venkateswarlu and D.J. Faulkner, Lamellarin alpha 20-sulfate, an inhibitor of HIV-1 integrase active against HIV-1 virus in cell culture, *J. Med. Chem.*, 1999, **42**, 1901–1907.
29. S.J. Teague and A.M. Davis, Hydrogen bonding, hydrophobic interactions, and the failure of the rigid receptor hypothesis, *Angew. Chem. Int. Edn.*, 1999, **38**, 736–749.
30. J.H. Bae, S. Alefelder, J.T. Kaiser, R. Friedrich, L. Moroder, R. Huber and N. Budisa, Incorporation of beta-selenolo[3,2-*b*]pyrrolylalanine into proteins for phase determination in protein X-ray crystallography, *J. Mol. Biol.*, 2001, **309**, 925–936.

31. N.E. Chayen, T.J. Boggon, A. Cassetta, A. Deacon, T. Gleichmann, J. Habash, S.J. Harrop, J.R. Helliwell, Y.P. Nieh, M.R. Peterson, J. Raftery, E.H. Snell, A. Hadener, A.C. Niemann, D.P. Siddons, V. Stojanoff, A.W. Thompson, T. Ursby and M. Wulff, Trends and challenges in experimental macromolecular crystallography, *Q. Rev. Biophys.*, 1996, **29**, 227–278.

32. J. Sedzik and U. Norinder, Statistical analysis and modelling of crystallization outcomes, *J. Appl. Crystallogr.*, 1997, **30**, 502–506.

33. S.W. Muchmore, J. Olson, R. Jones, J. Pan, M. Blum, J. Greer, S.M. Merrick, P. Magdalinos and V.L. Nienaber, Automated crystal mounting and data collection for protein crystallography, *Struct. Fold Des.*, 2000, **8**, R243–246.

34. J. Miao, K.O. Hodgson and D. Sayre, An approach to three-dimensional structures of biomolecules by using single-molecule diffraction images, *Proc. Natl. Acad. Sci. U.S.A.*, 2001, **98**, 6641–6615.

35. K. Palczewski, T. Kumasaka, T. Hori, C.A. Behnke, H. Motoshima, B.A. Fox, I. Le Trong, D.C. Teller, T. Okada, R.E. Stenkamp, M. Yamamoto and M. Miyano, Crystal structure of rhodopsin: A G protein-coupled receptor, *Science*, 2000, **289**, 739–745.

36. M.J. Sutcliffe, I. Haneef, D. Carney and T.L. Blundell, Knowledge based modelling of homologous proteins, Part I: Three-dimensional frameworks derived from the simultaneous superposition of multiple structures, *Protein Eng.*, 1987, **1**, 377–384.

37. A. Sali and T.L. Blundell, Comparative protein modelling by satisfaction of spatial restraints, *J. Mol. Biol.*, 1993, **234**, 779–815.

38. J.J. Hanak, The 'multiple sample concept' in materials research: synthesis, compositional analysis, and testing of entire multicomponent systems, *J. Mater. Sci.*, 1970, **5**, 964–971.

39. S.J. Teague, A.M. Davis, P.D. Leeson and T. Oprea, The design of leadlike combinatorial libraries, *Angew. Chem. Int. Ed. Engl.*, 1999, **38**, 3743–3748.

40. T.I. Oprea, A.M. Davis, S.J. Teague and P.D. Leeson, Is there a difference between leads and drugs? A historical perspective, *J. Chem. Inf. Comput. Sci.*, 2001, **41**, 1308–1315.

41. S.B. Shuker, P.J. Hajduk, R.P. Meadows and S.W. Fesik, Discovering high-affinity ligands for proteins: SAR by NMR, *Science*, 1996, **274**, 1531–1534.

42. H.J. Boehm, M. Boehringer, D. Bur, H. Gmuender, W. Huber, W. Klaus, D. Kostrewa, H. Kuehne, T. Luebbers, N. Meunier-Keller and F. Mueller, Novel inhibitors of DNA gyrase: 3D structure based biased needle screening, hit validation by biophysical methods, and 3D guided optimization. A promising alternative to random screening, *J. Med. Chem.*, 2000, **43**, 2664–2674.

43. B.K. Shoichet and I.D. Kuntz, Matching chemistry and shape in molecular docking, *Protein Eng.*, 1993, **6**, 223–232.

44. D.S. Goodsell, G.M. Morris and A.J. Olson, Docking of flexible ligands: Applications of AutoDock, *J. Mol. Recognition*, 1996, **9**, 1–5.

45. M. Rarey, B. Kramer, T. Lengauer and G. Klebe, Predicting receptor–ligand interactions by an incremental construction algorithm, *J. Mol. Biol.*, 1996, **261**, 470–489.

46. B.K. Shoichet, R.M. Stroud, D.V. Santi, I.D. Kuntz and K.M. Perry, Structure-based discovery of inhibitors of thymidylate synthase, *Science*, 1993, **259**, 1445–50.

47. P. Burkhard, P. Taylor and M.D. Walkinshaw, An example of a protein ligand found by database mining: description of the docking method and its verification by a 2.3 A X-ray structure of a thrombin-ligand complex. *J. Mol. Biol.*, 1998, **277**, 449–466.

48. J. Janin, Elusive affinities, *Proteins*, 1995, **21**, 30–39.

49. J. Tokarski and A.J. Hopfinger, Constructing protein models for ligand–receptor

binding thermodynamic simulations, *J. Chem. Inf. Comput. Sci.*, 1997, **37**, 779–791.

50. D. Horvath, A virtual screening approach applied to the search for trypanothione reductase inhibitors, *J. Med. Chem.*, 1997, **40**, 2412–2423.

51. A.C. Anderson, R.H. O'Neil, T.S. Surti and R.M. Stroud, Approaches to solving the rigid receptor problem by identifying a minimal set of flexible residues during ligand docking, *Chem. Biol.*, 2001, **8**, 445–457.

52. A.R. Leach, Ligand docking to proteins with discrete side-chain flexibility, *J. Mol. Biol.*, 1994, **235**, 345–356.

53. D.M. Lorber and B.K. Shoichet, Flexible ligand docking using conformational ensembles, *Protein Sci.*, 1998, **7**, 938–950.

54. H. Claussen, C. Buning, M. Rarey and T. Lengauer, FlexE: efficient molecular docking considering protein structure variations, *J. Mol. Biol.*, 2001, **308**, 377–395.

55. P.J. Goodford, A computational procedure for determining energetically favorable binding sites on biologically important macromolecules, *J. Med. Chem.*, 1985, **28**, 849–857.

56. C.M. Stultz and M. Karplus, MCSS functionality maps for a flexible protein, *Proteins*, 1999, **37**, 512–529.

57. G. Jones, P. Willett, R.C. Glen, A.R. Leach and R. Taylor, Development and validation of a genetic algorithm for flexible docking, *J. Mol. Biol.*, 1997, **267**, 727–748.

58. T.J. Ewing, S. Makino, A.G. Skillman and I.D. Kuntz, DOCK 4.0: search strategies for automated molecular docking of flexible molecule databases, *J. Comput. Aided Mol. Des.*, 2001, **15**, 411–428.

59. J. Aqvist, C. Medina and J.E. Samuelsson, A new method for predicting binding affinity in computer-aided drug design, *Protein Eng.*, 1994, **7**, 385–391.

60. S.S. So and M. Karplus, Comparative study of ligand–receptor complex binding affinity prediction methods based on glycogen phosphorylase inhibitors, *J. Comput. Aided Mol. Des.*, 1999, **13**, 243–258.

61. D. Rognan, S.L. Lauemoller, A. Holm, S. Buus and V. Tschinke, Predicting binding affinities of protein ligands from three-dimensional models: application to peptide binding to class I major histocompatibility proteins, *J. Med. Chem.*, 1999, **42**, 4650–4658.

62. A. Logean, A. Sette and D. Rognan, Customized versus universal scoring functions: application to class I MHC-peptide binding free energy predictions, *Bioorg. Med. Chem. Lett.*, 2001, **11**, 675–679.

63. A.R. Ortiz, M.T. Pisabarro, F. Gago and R.C. Wade, Prediction of drug binding affinities by comparative binding energy analysis, *J. Med. Chem.*, 1995, **38**, 2681–2691.

64. R.D. Head, M.L. Smythe, T.I. Oprea, C.L. Waller, S.M. Green and G.R. Marshall, VALIDATE – A new method for the receptor-based prediction of binding affinities of novel ligands, *J. Am. Chem. Soc.*, 1996, **118**, 3959–3969.

65. A. Vedani, P. Zbinden and J.P. Snyder, Pseudo-receptor modelling: a new concept for the three-dimensional construction of receptor binding sites, *J. Receptor Res.*, 1993, **13**, 163–177.

66. P. Zbinden, M. Dobler, G. Folkers and A. Vedani, PRGEN – pseudoreceptor modelling using receptor-mediated ligand alignment and pharmacophore equilibration, *Quant. Struct. Act. Relat.*, 1998, **17**, 122–130,

67. A.C. English, S.H. Done, L.S. Caves, C.R. Groom and R.E. Hubbard, Locating interaction sites on proteins: the crystal structure of thermolysin soaked in 2% to 100% isopropanol, *Proteins*, 1999, **37**, 628–640.

68. A.C. English, C.R. Groom and R.E. Hubbard, Experimental and computational mapping of the binding surface of a crystalline protein, *Protein Eng.*, 2001, **14**, 47–59.

69. D. Joseph-McCarthy, J.M. Hogle and M. Karplus, Use of the multiple copy simultaneous search (MCSS) method to design a new class of picornavirus capsid binding drugs, *Proteins*, 1997, **29**, 32–58.

70. C.M. Stultz and M. Karplus, MCSS functionality maps for a flexible protein, *Proteins*, 1999, **37**, 512–529.

71. A. Miranker and M. Karplus, Functionality maps of binding sites: a multiple copy simultaneous search method, *Proteins*, 1991, **11**, 29–34.

72. A. Miranker and M. Karplus, An automated method for dynamic ligand design, *Proteins*, 1995, **23**, 472–490.

73. C.M. Stultz and M. Karplus, Dynamic ligand design and combinatorial optimization: designing inhibitors to endothiapepsin, *Proteins*, 2000, **40**, 258–289.

74. V.J. Gillet, W. Newell, P. Mata, G. Myatt, S. Sike, Z. Zsoldos and A.P. Johnson, SPROUT: recent developments in the *de novo* design of molecules, *J. Chem. Inf. Comput. Sci.*, 1994, **34**, 207–217.

75. X.Q. Lewell, D.B. Judd, S.P. Watson and M.M. Hann, RECAP – retrosynthetic combinatorial analysis procedure: a powerful new technique for identifying privileged molecular fragments with useful applications in combinatorial chemistry, *J. Chem. Inf. Comput. Sci.*, 1998, **38**, 511–522.

76. A.L. Barabasi, V.W. Freeh, H. Jeong and J.B. Brockman, Parasitic computing, *Nature*, 2001, **412**, 894–897.

77. E. Overton, *Z. Phys. Chem.*, 1897, **22**, 189–209

78. C. Hansch, P.P. Maloney and T. Fujita, *Nature*, 1962, **178**, 4828.

79. M. Pastor, G. Cruciani, I. McLay, S. Pickett and S. Clementi, GRid-INdependent descriptors (GRIND): a novel class of alignment-independent three-dimensional molecular descriptors, *J. Med. Chem.*, 2000, **43**, 3233–3243.

80. J.A. Platts, M.H. Abraham, A. Hersey and D. Butina, Estimation of molecular linear free energy relationship descriptors. 4. Correlation and prediction of cell permeation, *Pharm. Res.*, 2000, **17**, 1013–1018.

81. C.A. Lipinski, Drug-like properties and the causes of poor solubility and poor permeability, *J. Pharmacol. Toxicol. Methods*, 2000, **44**, 235–249.

82. H. van de Waterbeemd, D.A. Smith and B.C. Jones, Lipophilicity in PK design: methyl, ethyl, futile, *J. Comput. Aided Mol. Des.*, 2001, **15**, 273–286.

83. K.G. Gurevich, Low doses of biologically active substances: effects, possible mechanisms, and features, *Cell Biol. Int.*, 2001, **25**, 475–484.

84. E.J. Calabrese and L.A. Baldwin, Hormesis: a generalizable and unifying hypothesis, *Crit. Rev. Toxicol.*, 2001, **31**, 353–424.

85. K. Linde, M. Hondras, A. Vickers, Gt.G. Riet and D. Melchart, Systematic reviews of complementary therapies – an annotated bibliography. Part 3: Homeopathy, *BMC Complement. Altern. Med.*, 2001, **1**, 4–11.

86. D.R. Flower, Modelling G-protein-coupled receptors for drug design, *Biochim. Biophys. Acta*, 1999, **1422**, 207–234.

87. D. Gillies, *Artificial Intelligence and the Scientific Method*, Oxford University Press, Oxford, 1996.

High-Throughput X-Ray Crystallography for Drug Discovery

Tom L. Blundell,[1, 3] Chris Abell,[2, 3] Anne Cleasby,[3] Michael J. Hartshorn,[3] Ian J. Tickle,[3] Emilio Parasini[1, 3] and Harren Jhoti[3]

[1]DEPARTMENT OF BIOCHEMISTRY, UNIVERSIY OF CAMBRIDGE, TENNIS COURT ROAD, CAMBRIDGE CB2 1GA, UK
[2]UNIVERSITY CHEMICAL LABORATORY, LENSFIELD ROAD, CAMBRIDGE CB2 1EW, UK
[3]ASTEX TECHNOLOGY, 250 CAMBRIDGE SCIENCE PARK, MILTON ROAD, CAMBRIDGE CB4 0WE, UK

1 Introduction

Drug discovery remains a risky business. Most lead compounds – those that might provide useful drugs – do not make it to clinical trials. Less than 10% of drug candidates survive safety and clinical evaluation, and two out of three marketed products do not yield a positive return on investment.[1] Furthermore, attrition rates of drug candidates are likely to increase, since the industry has little experience of the potential toxicity of novel classes of drugs that may be suggested by the new genome-derived targets.

Until a few years ago many leads were derived from plant products, many of which had proved to be efficacious in traditional medicines. Others were identified by screening fermentation broths. As knowledge increased, ligands could be identified from the biochemistry of their targets. In the eighties the industry began to use knowledge of the structures of both ligands and target proteins to suggest leads and then to optimise their binding to the target proteins.[2] X-Ray crystallography, NMR spectroscopy and molecular modelling became widely used tools to guide the medicinal chemist. Structural information helped the synthetic chemist to optimise lead compounds by building better interactions with the protein, resulting in improved potency and selectivity.[3] Indeed, there are now several drugs on the market that originated from this structure-based design approach. The most commonly cited are HIV drugs such as Agenerase and Viracept that were developed using the crystal structure of HIV proteinase.[4]

Due to the mounting pressures to increase productivity, pharmaceutical companies have been constantly looking for technology-driven solutions. In the 1990s the focus of discovery moved to 'diversity-based' screening, involving random high-throughput screening (HTS) of compound libraries, many of them synthesised using combinatorial chemistry. However, this approach has had less impact on the number of new chemical entities discovered than was originally hoped.[5] The emphasis now is to combine this approach with more rational approaches, for example by using 'knowledge-based' or 'focused' screening. Here, the knowledge of the biochemistry and/or structure is used to pre-select compounds or fragments that are predicted to bind, often using *in silico* screening methods. These compounds are then tested in bioassay screens, often at higher concentrations than normal. Initial data on the number of compounds that need to be screened for each tractable lead identified show that focused medium/low throughput screening is more effective than random high-throughput screening.

There is now growing interest in applying biophysical techniques to lead discovery. Applications of mass spectrometry, isothermal calorimetry, NMR spectroscopy and X-ray crystallography to lead discovery have been recently described.[6-10] A key advantage of these biophysical methods over traditional bioassays for lead discovery is their ability to detect the binding of relatively low affinity compounds. Most bioassays performed in HTS formats are designed only to detect compounds that show potency better than $10\,\mu M$. Thus, many weak, but chemically tractable, leads may be missed by the HTS campaigns being performed in most pharmaceutical companies.

We are developing a structural approach to lead discovery that can detect weak leads and optimise them into useful drug candidates.[11] In this approach, which we call structural screening, we couple virtual screening of compounds with rapid X-ray crystallographic analysis of protein/ligand complexes (Figure 1). To establish structural screening as a useful industrial lead discovery approach we are developing methods for performing X-ray crystallography in a high throughput and automated manner. In this paper we describe some pilot experiments using trypsin that illustrate the application of structural screening to chemical lead generation.

2 Methods

High throughput analysis of protein/ligand complexes requires automation of data collection and analysis of series of isomorphous crystals. We are, therefore, less concerned with crystallographic phase determination, and more with calculation and interpretation of difference Fouriers in order to position ligands in a previously defined crystal structure. The first step is, therefore, to select a chemical library for soaking experiments. The second is to arrange for automated high throughput soaking of cocktails of these compounds into crystals. We then need to collect X-ray data efficiently and quickly, and automate the interpretation of the difference electron density maps, so that we can both recognise the bound ligand and define its position in the complex.

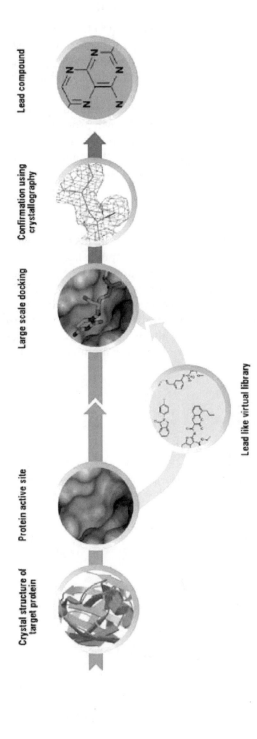

Figure 1 *Structural screening: this approach couples virtual screening with high throughput X-ray crystallography of protein/ligand complexes. The virtual screening step may involve selection based on chemical similarity, a pharmacophore and/or large-scale docking into a protein active site. Compounds identified from the virtual screening step are then used in rapid X-ray crystallographic analysis using AutoSolve® to define experimentally their binding modes*

2.1 Compound Selection for Crystal Soaking

A relatively small library of molecular fragments (300–500 compounds) can sample a significant amount of chemical space. Therefore, screening small-molecule fragments can offer a very effective and elegant way to identify novel pharmacophores for a new target protein, so avoiding synthesis of huge combinatorial libraries.

Two sets of small-molecule fragments can be generated for initial structural screens. The first is a focused set, chosen using known protein binders as starting points for chemical similarity searches. The second set is a 'universal fragment set' chosen by combining known drug scaffolds with commonly found drug side chains. These were developed for our target protein trypsin on the basis of previous published work.

2.1.1 Focused Set. Earlier crystallographic experiments had shown the binding of three small-molecule fragments to trypsin (unpublished results). These were benzamidine, 4-aminopyridine and cyclohexylamine. These molecules were each used as starting points for similarity searches of chemical databases. A further three molecules, thought capable of making similar interactions, were also used as starting points. These molecules were histamine, 2-aminoimidazole and 4-aminoimidazole. Finally, proflavin was also included, as this has been observed to bind to a related target.[12] The set of seven compounds used for similarity searches is shown in Figure 2.

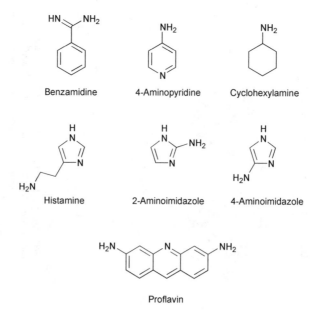

Figure 2 *Compounds used as starting points for similarity searches*

The similarity searches were performed on the Available Chemicals Directory (version acd992) using Merlin Command Language from Daylight (Daylight Chemical Information Systems, Los Altos, USA). Using the Daylight Similarity Metric, compounds were chosen with a score of greater than 0.7, where a score of one would indicate identical compounds. The similarity searches generated a total of 224 compounds. The set of compounds identified from the seven starting points was then processed to remove compounds that contained reactive or functional groups associated with toxicity. The resulting compounds obtained by this process were then examined for cost, availability and subjective interest and a subset chosen.

2.1.2 Universal Fragment Set. Simple organic ring systems that are found in many drug molecules can be considered as low molecular weight frameworks. Decorating these frameworks with side chains containing the most frequently found functional groups in drugs can generate a Universal fragment set. The compounds are generated as SMILES strings, which are then searched for in a database of available compounds.[13] A total of 4513 compounds were generated in the virtual enumeration stage for trypsin of which 401 were available from UK chemical suppliers. Of these we obtained 353 compounds for structural screening against trypsin. The compounds were prepared for soaking experiments by dissolving into DMSO or water. The final concentration of each compound in a soaking experiment was 25 mM.

2.2 X-Ray Data Collection

Crystals of trypsin were grown by adapting a published procedure;[14] 5.4 mg of bovine trypsin were dissolved in 100 μl of 50 mM sodium acetate, pH 5.6, 18 mM $CaCl_2$, 0.81 M ammonium sulfate, 5 mg ml^{-1} benzamidine and equilibrated in hanging drops over wells containing 1.6 M ammonium sulfate. The resulting crystals were repeatedly backsoaked to remove any bound benzamidine before soaking. The soaks with single crystals were done at a concentration of 100 mM, the cocktails at a concentration of 25 mM (in 8 compound mixes) for 1 hour. The crystals were then transferred into cryoprotectant before data collection at 100 K.

X-Ray data were collected from trypsin crystals using a Jupiter CCD with a Rigaku RU H3R generator and processed using D*trek.[15]

2.3 Interpretation of Difference Maps using AutoSolve®

All X-ray data were automatically analysed and electron density interpreted using AutoSolve®. Examples of small-molecule fragments bound to trypsin are shown in Figure 3.

One of the key bottlenecks in performing high throughput X-ray crystallography of protein/ligand structures is the time and skills required for interpretation and analysis of X-ray data. Once the structure of the target protein is known,

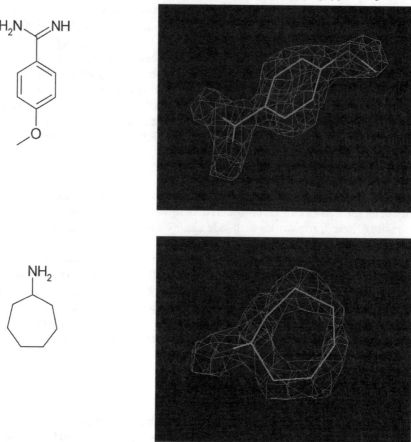

Figure 3 *Examples of small-molecule fragments bound into a pocket of trypsin. The electron density was interpreted and models of compounds automatically fitted using AutoSolve®. The electron density maps are contoured at 3σ and density due to protein and solvent has been removed for clarity*

there is a need to complex many different compounds to the target and to establish rapidly their binding modes. Conventionally, this requires an experienced X-ray crystallographer to interpret and analyse each X-ray data set collected from a crystal where the protein has been complexed with a compound either by co-crystallisation or by a soaking experiment. To accelerate this stage we have developed AutoSolve®,[16] which allows the rapid structure solution of protein/ligand complexes by interpreting and analysing the X-ray data without the need for manual intervention. In each case AutoSolve® was able to identify the bound fragment from a cocktail of fragments that was used in the soaking experiment.

3 Results

Crystallographic experiments involving soaking of crystals with cocktails of compounds have been previously reported.[10] However, we have found that manual interpretation of electron density from soaking experiments using cocktails of compounds is often difficult and can be unreliable. It is only by employing a reliable and robust data analysis tool such as AutoSolve® that an objective interpretation can be reached. Figure 3 shows examples of clear electron density that were unambiguously interpreted by AutoSolve®; in each case the binding mode of the small-molecule fragment is clearly defined. It is worth noting that even though the binding affinity of these small-molecule fragments is expected to be in the millimolar range the binding mode is specific and the key interactions are clearly defined. This data indicates that the weak binding of small-molecule fragments can be detected using X-ray crystallography.

In this study we have shown how virtual screening coupled with high throughput X-ray crystallography can be used to screen compound libraries consisting of hundreds of small-molecule fragments. The detection and subsequent optimisation of these weak binding small-molecule fragments could be the basis of a powerful approach for lead discovery resulting in novel drug candidates.

References

1. S.F. Campbell, *Clin. Scie.*, 2000, **99**, 255.
2. P.J. Whittle and T.L. Blundell, *Ann. Rev. Biophys. Biomol. Struct.*, 1994, **23**, 349.
3. J. Greer, J.W. Erickson, J.J. Baldwin and M.D. Varney, *J. Med. Chem.*, 1994, **37**, 1035.
4. T.L. Blundell, *Nature*, 1996, **384S**, 23.
5. A. R. Leach and M. M. Hann, *Drug Discovery Today*, 2000, **5**, 326.
6. F.J. Moy, K.Haraki, D. Mobilio, G. Walker, R. Powers, K. Tabei, H. Tong and M.M. Siegel. *Anal. Chem.*, 2001, **73**, 571.
7. W.H.J. Ward and G.A. Holdgate, 'Progress in Medicinal Chemistry', Elsevier Science, Vol. 38, Chapter 7, p. 309.
8. P.J. Hajduk, M. Bures, J. Praestgaard and S.W. Fesik, *J. Med. Chem.*, 2000, **43**, 3443.
9. J. Fejzo, C.A. Lepre, J.W. Peng, G.W. Bemis, Ajay, M.A. Murcko and J.M. Moore, *Chem. Biol.*, 1999, **6**, 755.
10. V.L Nienaber, P.L. Richardson, V. Klighofer, J.J. Bouska, V.L. Giranda and J. Greer, *Nat. Biotechnol.*, 2000, **18**, 1105.
11. T.L. Blundell, H. Jhoti and C. Abell, *Naure Rev. Drug. Discovery*, 2002, **1**, 45.
12. E. Conti E, C. Rivetti, A.Wonacott and P. Brick, *FEBS Lett.*, 1998, **425**, 229.
13. D. Weininger, *J. Chem. Inf. Comp. Sci.*, 1988, **28**, 31.
14. G. Zhu, Q. Huang, Z. Wang, M. Qian, Y. Jia and Y. Tang, *Biochim. Biophys. Acta*, 1998, **1429**, 142.
15. J.W. Pflugrath, *Acta Cryst. D*, 1999, **55**, 1718.
16. Patent pending.

Trawling the Genome for G Protein-coupled Receptors: the Importance of Integrating Bioinformatic Approaches

Teresa K. Attwood[1] and Darren R. Flower[2]

[1]SCHOOL OF BIOLOGICAL SCIENCES, THE UNIVERSITY OF MANCHESTER, OXFORD ROAD, MANCHESTER M13 9PT, UK
[2]EDWARD JENNER INSTITUTE FOR VACCINE RESEARCH, COMPTON, NEWBURY, BERKSHIRE RG20 7NN, UK

1 Introduction

G protein-coupled receptors, or GPCRs, also known as 7TM, heptahelical, or serpentine receptors, form one of the largest, most diverse, yet best studied groups of cell-surface molecules. GPCRs mediate an enormous range of key physiological processes, including the perception of light, taste, smell, and pain. Because of the breadth and importance of these biological roles, as attested with receptor knockouts and their link to hereditary diseases, members of the GPCR family have become important pharmacological target molecules. Indeed, over 50% of all marketed drugs act at GPCRs – including a quarter of the 100 top-selling drugs – yielding sales of over 16 billion US dollars per annum (see Table 1). Yet many therapies involving such drugs have efficacy problems and limiting side effects, because the compounds do not properly differentiate between receptor subtypes. There is therefore considerable interest from both clinicians and pharmaceutical companies in developing therapeutic specificity by identifying the single receptor subtype responsible for mediating a particular pathophysiology, and thereby defining an appropriate intervention point. Ultimately, the aim is to design drugs that eliminate, or reduce, unwanted effects, while still conferring the desired therapeutic benefit. For example, muscarinic agonists, especially those that activate the M_1 receptor subtype, have been considered potentially useful in treating Alzheimer's disease – it was thought that the cardiovascular and gastrointestinal side effects associated with non-specific muscarinic agents could be avoidable, as the M_1 receptor is found in the brain and may be involved with cognition, while other subtypes regulate heart and gastrointestinal functions. Hand in hand with the objective of tailoring subtype

Table 1 *Marketed drugs targeted at GPCRs. Marketed drugs which act* via *GPCRs drawn from the 100 top selling pharmaceuticals worldwide during 1997. Total combined sales for these compounds exceeds 15845 million $ for 1997. The drugs are listed alphabetically, and are not ranked by sales. Data taken from [ref. 4]*

Drug	Commercial name	Activity
Atenolol	Tenormin	β_2 antagonist
Buspirone	Buspar	$5HT_{1a}$ agonist
Cetirizine	Zyrtec	antihistamine H_1 antagonist
Cimetidine	Tagamet	H_2 antagonist
Cisapride	Prepulsid	$5HT_4$ ligand
Doxazosin	Cardura	α_1 antagonist
Famotidine	Gaster	H_2 antagonist
Goserelin	Zoladex	LHRH agonist
Ipratropium	Atrovent	Mixed Muscarinic antagonist
Metoprolol	Betaloc	β_1 antagonist
Nizatidine	Axid	H_2 antagonist
Leuprolide	Lupron	LHRH agonist
Leuprorelin	Prostap Sr	LHRH agonist
Loratadine	Claritin	antihistamine H_1 antagonist
Losartan	Cozaar	AT_1 antagonist
Olanzapine	Zyprexa	Mixed $D_2/D_1/5HT_2$ antagonist
Ranitidine	Zantac	H_2 antagonist
Risperidone	Risperdal	Mixed $D_2/5HT_2$ antagonist
Salbutamol	Ventolin	β_2 agonist
Salmeterol	Serevent	β_2 agonist
Sumatriptan	Imigran	$5HT_1$ agonist
Terazosin	Heitrin	α_1 antagonist

specificity is the desire to achieve orally active compounds, or drugs, with enhanced duration of action and improved ADME properties, as well as an increasing interest in the combination of clearly defined potencies at several receptor types within the same molecule.[1–3]

GPCRs provide an excellent illustration of the phenomenon described by Jacob as 'molecular tinkering':[5,6] these proteins have been very successful in evolution, successfully adapting a common structural framework to fulfil innumerable different functions.[7] Amongst these, GPCRs mediate chemotaxis, stimulation and regulation of mitosis, and the opportunistic entry of viruses into cells.[8,9] Such functional diversity is achieved *via* interactions with a wide variety of ligands, including peptides, glycoproteins, small molecule messenger molecules, such as adrenalin, and even photons, as well as through the diversity of second messenger systems they activate. GPCRs derive their name from their interaction with heterotrimeric G proteins, but it has now become clear that they also interact with a wide variety of other intracellular molecules.[9,10] For example, the adaptor molecule arrestin couples GPCRs to the activation of Src-like kinases and facilitates the formation of multimolecular complexes. Other structural components (*e.g.*, polyproline-containing regions, PDZ, SH2 and SH3 domains) mediate direct interactions between GPCRs and a variety of intracellular signalling molecules.[8,10] The extraordinary richness of mechanisms by which GPCRs transduce signals is only now being appreciated, and the conventional picture of classical second-messenger-generating systems, operating *via* single biochemical routes, is changing rapidly in favour of an integrated view, involving an intricate network of cytoplasmic signalling pathways.

A pivotal achievement of recent times was the public release of the draft human genome, which promises to improve our understanding of diverse aspects of biology, yielding a healthier future with safer, more 'personalised' medicines. To meet these promises, the information sequestered within genomes needs to be extricated. In particular, researchers need rapid, easy-to-use, *reliable* tools for functional characterisation of raw sequence data. Previous estimates of the number of GPCRs encoded by the human genome suggested that they represented about 1 percent of all genes, with another 1,000–2,000 GPCRs involved in olfaction. The current estimate of the size of the human genome has been revised down from a figure in excess of 100,000 to an initial estimate of 35,000–40,000 genes. Other, more recent, 'best guesses' place the number nearer 65,000–75,000. Based on the draft human genome, subsequent analysis, using an automated pipeline based on a combination of BLAST and hidden Markov models, has suggested a total of about 900 rhodopsin-like GPCRs, of which about 420 are involved in olfaction and about 60 are novel orphan receptors, many with unassigned functions. Even amongst those receptors whose natural, endogenous ligands are known, there are still many receptors for which synthetic agonists or antagonists are lacking. Consequently, there is still much untapped potential for GPCR-based drug discovery. GPCR-orientated research is a ubiquitous aspect of target identification and drug design programmes within most major pharmaceutical companies and will remain of interest for as long as the human genome holds undiscovered receptors that may present new therapeutic targets. Within this context, we describe here some of the *in silico* approaches by means of which genome data may now be profitably analysed.

2 *In silico* Tools for Sequence Analysis

Today, the number of available biological databanks is now legion – they require a database of their own just to catalogue them. Dbcat, for example, lists over 500.[11] Rationalising the vast quantity of data emerging from innumerable genome projects has required both an unprecedented level of global co-operation and an ever increasing degree of automation in data handling and analysis. However, automation can carry a heavy price. In the field of genomics, for example, software 'robots' are used in the process of functional annotation of newly-determined sequences, but they pose a threat to information quality because they can introduce and propagate mis-annotations.[12] Although curators are aware of this problem, and strive to reduce errors, nonetheless databases are historical products and users should therefore always bear in mind that they are imperfect.

The first step towards functional characterisation of a new sequence usually involves searching a sequence database with tools such as FastA[13] or BLAST.[14] Such searches can reveal obvious similarities between the query sequence and a range of sequences in the database. The difficulty then lies in the reliable inference of homology (the verification of a divergent evolutionary relationship) and, from this, the inference of biological function. In an ideal world, a search output will show unequivocal similarity to a well-characterised protein over the full length of the query. More commonly, the output will reveal no significant hits, and the usual scenario lies between the extremes, with a list of partial matches to diverse proteins, many of them uncharacterised, and some with dubious or contradictory annotations.[15]

There are various reasons for this confusion. First, the greatly enlarged size of modern sequence databases, and their population by increasing numbers of poor quality partial sequences, gives rise to a greater likelihood that relatively high-scoring, but coincidental, matches will be made to a query. Secondly, issues arise from the existence of multi-gene families because database search techniques cannot differentiate between a matched paralogue (a homologue that performs different but related functions within the same organism) and a matched ortho-logue (the functional counterpart of a sequence in another species). Thirdly, if not masked, low-complexity matches may interfere with search outputs. The modu-lar/domain nature of many proteins may also be problematic, as it may not be clear, when making matches to multi-domain proteins, which domain or do-mains correctly correspond to the query. Second, even if the right domain has been found, it may not be appropriate to transfer the functional annotation to the query because the function of the matched domain may be different, depend-ing on its precise biological context.

Achieving consistent, reliable functional assignments can prove to be a com-plex problem. As a result, in addition to routine searches of the sequence databases, it is now usual to extend search strategies to include a range of protein families, usually encoded in what are commonly referred to as motif databases. These databases distil information within groups of related sequences into potent descriptors or discriminators that can greatly assist family diagnosis. Searching pattern databases can be both more sensitive and more selective than sequence database searching because derived family discriminators can detect weaker regions of similarity, and they can exploit differences between sequences as well as their similarities, as we shall see later. Different analytical approaches have been used to create a bewildering array of discriminators: regular expres-sions, rules, profiles, signatures, fingerprints, blocks, *etc.*[16] The different descrip-tors have different diagnostic strengths and weaknesses, and different areas of optimum application, and have been used to create different pattern databases, which can vary in their composition. It is therefore important to know how best to exploit them.

3 Protein Motif Databases

Multiple sequence alignment lies at the heart of the analysis methods that underpin pattern databases. When an alignment is created, more and more distantly related sequences can be included, requiring insertions and deletions to bring the equivalent parts of sequences into the correct register. As a result of this process, distinct islands of conservation emerge from an ocean of mutational change. These conserved regions generally correspond to the core structural elements of the protein and are usually termed motifs. Different techniques have emerged that can exploit this encoded conservation: they fall into three categories, depending on whether they use full domain alignments, multiple or single motifs. All such methods require the derivation of a discriminatory representation of the conserved elements of the alignment, providing a signature characteristic of the family, which, in turn, can be used to facilitate diagnosis of future sequences.

The first pattern database to have been created was PROSITE,[17] which uses regular expressions to encode single motifs. A single motif is often not sufficient to capture the full extent of a protein family, and hence additional regular expressions may need to be derived. PROSITE patterns are deterministic: when matching such patterns, a query sequence must match the expression exactly, or it will not be diagnosed – the approach will not tolerate any residue mismatch. Conversely, many otherwise unrelated sequences will match these patterns exactly. Together, these features complicate genome analysis because a match to a regular expression is not necessarily true and a mis-match is not necessarily false.

To overcome this diagnostic limitation, a multiple-motif approach was devised. Here, motifs are encoded as residue frequency matrices, without the use of mutation or substitution data to weight the matches. This is the basis of the fingerprint approach, which underpins the PRINTS database.[18] A related technique exploits substitution matrices to score motif matches, and this weighted-motif approach is the basis of the Blocks database.[19] Finally, to take advantage of the gapped regions between motifs, which provide important information about inter-motif distances, approaches were devised to encompass the full length of conserved domains. Examples include profiles (which use an absolute scoring system that exploits evolutionary weights and differential gap penalties to prevent their occurrence in core secondary structure elements) and hidden Markov models, or HMMs (which use a probabilistic approach to assign match, delete and insert states to all positions in an alignment). These methods underpin the Profiles[17] and Pfam[20] databases.

Understanding the relative strengths and weaknesses of these approaches, gathering the range of different outputs they provide, and arriving at some sort of consensus view of their results, is often challenging. In an effort to make the process more straightforward and to provide a single resource for sequence analysis, the curators of PROSITE, Profiles, PRINTS and Pfam have created a unified database of protein families, termed InterPro.[21] This database is an integrated family annotation resource, based primarily on existing documentation in PROSITE and PRINTS. Subsequent releases have also included data from ProDom and SMART. InterPro is an important development because the participating databases now use a common nomenclature, and share consistent, standardised documentation for protein families, domains and functional sites. Potentially this development will greatly facilitate the correct inference of function by gathering consensus evidence, using a common language, in order to allow users to pinpoint homologous relationships with greater confidence.

4 The *In silico* Identification of Specific Receptor Subtypes

Most computational, or *in silico*, strategies for identifying GPCRs are still based on similarity searches using primary database search tools, such as BLAST. GPCRs are, phyletically speaking, widely distributed in eukaryotic organisms. Several families of GPCRs are distinct at the protein sequence level. By far the most extensive of these, the so-called rhodopsin-like family, currently has 3000–4000 sequences available in public databases. Given this large set of known GPCR sequences, characterising family members should, in principle, become more straightforward. However, as outlined above, the growth of noise in the source databases has meant that it is actually becoming more and more difficult to identify and classify members of this large superfamily in a reliable way. For example, it is apparent that BLAST 'sees' similarity between pairs of sequences in a different way when compared with family-based approaches. It reveals generic similarities (*e.g.*, it can show that the sequences being compared share several hydrophobic regions) but it cannot recognise individual family traits (*i.e.*, it cannot distinguish the differences between the sequences, such as specific ligand-binding motifs). Similarly, most pattern databases tend to provide generic signatures that are only capable of diagnosing superfamily relationships. Thus, these databases might recognise that a sequence belongs to the rhodopsin-like GPCR superfamily, but they cannot offer insights into the particular family to which it belongs. But it is no longer sufficient to say that a newly determined sequence is a GPCR – *i.e.*, one of possibly 50 sub-families. Ideally, we would wish to identify the specific receptor family and sub-family to which it belongs, and to begin saying something meaningful about ligand binding. For drug discovery scientists interested in the treatment of obesity, for example, who might specifically wish to identify type 4 melanocortin receptors (which are important in regulating appetite), a superfamily-level diagnosis is of limited value.

To facilitate the identification of particular subtypes, a systematic analysis of GPCRs has been undertaken as part of the effort to populate the PRINTS database. In an early attempt to identify GPCRs in sequence databases, a diagnostic fingerprint for rhodopsin-like GPCRs was developed based on common patterns of conservation in the seven transmembrane (TM) regions.[22] This operated well when publicly available GPCR sequences numbered less than 100. But, as the size of the superfamily and the source databases have both grown dramatically, the diagnostic performance of the fingerprint has deteriorated.

Although, in general, key regions of these sequences are well conserved, characteristic variations have been observed in the 7TM signatures, such that many obvious family members do not now share all features of the fingerprint. Thus, more recently, efforts were directed toward characterising not just the superfamilies, but also their constituent families and subtypes. To this end, sequence alignments were created manually for each level of the family hierarchy. Regions of similarity and differences between alignments were located and used to build a range of discriminatory 'fingerprints'. As discussed above, fingerprints are groups of motifs that together provide a signature of family membership. Motifs tend to reflect functionally or structurally important regions (*e.g.*, TM domains, protein–protein interaction sites, ligand-binding sites, and so on), thereby characterising the families in which they are found. In this analysis, within superfamilies, the motifs encode the only features common to all members (*i.e.*, the scaffold of 7TM domains). By contrast, at the family level, the motifs focus on those regions that characterise the particular family, but distinguish it from the parent superfamily; predictably, these are usually small parts of TM and loop regions. For receptor subtypes, the distinguishing traits are largely present in the N- and C-terminal regions, and in the third cytoplasmic loop.

This has led to the development of a GPCR-specific subset of the PRINTS database. The essential difference between this and other web-based GPCR databases is its emphasis on sequence diagnosis. The intention is to go beyond the look-up table approach, with a view to providing interactive analytical tools for identification and characterisation of family members at the sequence level. In essence, this GPCR pattern-recognition resource provides a finely-tuned diagnostic tool for GPCR sequence recognition. To date, more than 200 fingerprints have been created that distinguish GPCRs at the levels of superfamily, family and specific receptor subtype.[23] For a given query, it is thus possible to determine to which GPCR superfamily the sequence belongs (*e.g.*, whether rhodopsin-like, secretin-like, *etc.*); of which family it is a member (*e.g.*, whether muscarinic, adrenergic, *etc.*); and which subtype its sequence signature most resembles (*e.g.*, whether M_1, M_2, M_3, *etc.*).

An interesting perspective on this result can be achieved by using the graphical output option from InterPro's sequence search facility. Results are plotted for each of the constituent databases, from which it is possible to place the fingerprint matches in context and see at a glance which regions of a sequence are matched by PRINTS, PROSITE, Profiles and Pfam. This example illustrates the fine-tuning that fingerprints add to the diagnostic process, being the only resource to offer family- and sub-type-specific diagnoses, and for which the matched 'blobs' have any significant meaning – *i.e.*, whether TM domains of the superfamily scaffold, N- and C-termini, or specific loop regions (see Figure 1).

5 Structural Data

There are now well in excess of 2,000 structurally distinct, solved protein and peptide structures available in the public domain. These are largely soluble,

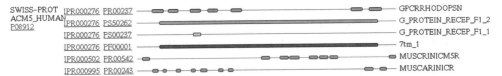

Figure 1 *InterPro graphical output for the GPCR sequence ACM5_HUMAN. The depic-*
ted 'blobs' correspond to hits to different discriminators. The first 4 lines show hits
to the superfamily discriminators in PRINTS (showing each of the 7TM motifs),
PROSITE profiles, PROSITE patterns and Pfam respectively. The final two lines
show hits to the family- and subtype-specific discriminators from PRINTS [show-
ing characteristic motifs in the N- and C-terminal and loop regions (especially the
third cytoplasmic loop, potentially important in G protein coupling), and parts of
the TM domains (most likely involved in ligand binding)]

globular proteins. Where once each structure was a landmark, it now takes a
remarkable protein to stand out from the crowd. The recently published structure of rhodopsin, determined to 2.8 Å resolution, is such a protein.[24] As the first
GPCR structure to be solved, and regarded by many as the archetypal GPCR,
rhodopsin has profound implications for the future bioinformatic discovery and
analysis of members of this superfamily. Hitherto, and as a consequence of their
preeminence in pharmaceutical research, much work has gone into the
modelling of GPCRs. A structurally related protein from purple bacteria, bac-
teriorhodpsin (BR) was first determined in 1990 using electron crystallography,[25]
and a completed X-ray structure was published in 1997.[26] However, BR bears no
sequence relationship to rhodopsin or other GPCRs, but nonetheless hundreds
of models were based upon it. More recently, theoretical models based on low
resolution and other experimental data have been proposed, again leading to a
flurry of derivative models. With the structure of rhodopsin, such models are
now more or less redundant. However, some of the data upon which such models
were built remains valid, particularly mutagenesis data, which can implicate the
role of particular residues in the interaction between ligand and receptor. Such
data helps us understand the conformational differences between receptor types.
Our understanding of other all-helical proteins, such as the globins or P450s,
suggests that while the topology of interacting helices is conserved, there can be
significant shifts in the relative positions of equivalent helices. So, while the
overall shape of a GPCR will be well maintained, the exact disposition of one
helix relative to another will vary from receptor to receptor, suggesting that
mutagenesis results will still greatly inform future homology modelling.
While the over-expression, purification, and crystallisation of membrane proteins remain difficult technical obstacles, it is interesting to note that, twenty
years ago, solving the structure of a soluble protein was still a relatively rare and
significant event, the number of skilled macromolecular crystallographers was
limited, and the number of crystallographic laboratories was small. Today,
several crystal structures are solved each and every day in one of the several
hundred macromolecular crystallography laboratories around the world staffed
by a huge community of trained crystallographers. Notwithstanding the capri-

cious nature of protein crystallisation, structure solution has become almost commonplace. As remaining technical problems are solved, with the necessary skills becoming more widespread, the study of GPCRs will become similarly routine, and we may look forward to significant advances in our understanding of their structure. This is certainly the hope of the semi-academic MEPNET initiative [http://www.mepnet.org/], as well as many other pharmaceutical, biotech, and start up companies, which are all working to this end.

At the same time, growth in our understanding of GCPR structure will allow much more detailed bioinformatic analysis. The correlation of biological function and phylogeny with individual sequence variation,[27,28] useful in itself as a means of identifying important functional residues, will become much more useful when placed in a structural context. The ability to perform full atomistic simulations of drug GPCR interactions, while still some way away, will likewise inform both bioinformaticians and molecular modellers.

6 Concluding Remarks

The technologies of target identification – genomics and proteomics – are now delivering an unprecedented volume of new genes and gene products for evaluation within pre-clinical research. Many of these targets may well be found to be members of the GPCR family. While the characterisation of orphan receptors may provide a source of novel targets well into the future,[29] clinicians and medicinal chemists remain interested in the development of new drugs targeted against well known and well investigated GPCRs. Within the integrated research environment that characterises the modern pharmaceutical industry, target finding and validation is the fountainhead from which all novel drug discovery projects will flow. Bioinformatics is a key component of this endeavour.

BLAST and FastA have been the mainstays of bioinformatic genome annotation efforts because they are simple both to use and to implement. However, their 'facile' use has led to problems as 'top hits' have often been used to transfer functional annotation from matched sequences to query sequences. Identifying similarity relationships between sequences is clearly not the same as identifying their functions, and failure to appreciate this fundamental point has generated and propagated annotation errors, and problems of all kinds for users of today's databases.

GPCR fingerprints allow more specific and reliable diagnoses than pairwise methods, yielding information from the level of the superfamily down to the individual receptor subtype. No other computational approach currently offers such a hierarchical discriminatory system for this important class of receptors. The fingerprint resource is thus a valuable complement to family and domain databases such as PROSITE and Pfam, offering potent diagnostic opportunities that have not been realised by other pattern-recognition methods. Fingerprint selectivity offers new opportunities to explore correlations between specific motifs and ligand binding or G protein coupling. With the availability of the first draft of the human genome, this collection of diagnostic GPCR fingerprints promises to facilitate both the identification of potential new drug targets and computational strategies to characterise orphan receptors.

Used together, pairwise- and family-based search tools offer the best means of mining the human genome for novel receptors. Preliminary results using a combination of such approaches have revealed that the total number of sensory and non-sensory GPCRs in the human genome is likely to be smaller than expected: lying somewhere between 700 and 900.[30,31] Future challenges in the elucidation of new therapeutic opportunities lie in unravelling the many non-G protein-coupled cellular pathways used by GPCRs, and also in determining the roles of alternative-splicing, receptor oligomerisation, and association of 7TM units with accessory proteins in receptor function. Moreover, some human diseases are associated with rare GPCR mutations and it is possible that widely distributed polymorphisms in GPCR genes may allow selective therapeutic strategies for population subgroups through the development of pharmacogenetics.

In this review, we have tried to emphasise the need for concerted, integrated protocols that highlight, in turn, the different perspectives offered by fundamentally different, yet ultimately complementary, methods of sequence analysis. For example, BLAST and FastA have a key place within bioinformatics, offering, as they do, broad brush strokes. Fingerprinting, and other motif based search methods, adds the fine detail. Structural modelling offers yet another perspective, increasing the nominal resolution to the atomic scale, and allowing us to explore the important physico-chemical properties that underlie drug binding. All of these different views are important. Together they will provide much more informative, much more detailed pictures of GPCR structure and function. They should also provide crucial help in the all-important process of target validation, by maximising the information available for a given target and minimising the risk associated with pursuing the less promising candidates.

Many GPCRs are orphan receptors with, as yet, no identified ligand, but as functional genomics begins to elucidate their physiological role, new therapeutic opportunities will follow. As pressure mounts on the pharmaceutical industry to shorten timescales and increase its cost-effectiveness, bioinformatic analysis is becoming more and more important. Target finding and validation has broad influence affecting many upstream functions. Of all the targets to be found, GPCRs remain one of the most important. The success of GPCR target discovery will continue to maintain the pre-eminent position of this family with pharmaceutical research for some time to come, promising much for the future.

Acknowledgements

TKA is a Royal Society University Research Fellow.

References

1. Y. Sasaki and T. Chiba, Novel deltorphin heptapeptide analogs with potent delta

agonist, delta antagonist, or mixed mu antagonist/delta agonist properties, *J. Med. Chem.*, 1995, **38**, 3995–3999.

2. H.S. Jae, M. Winn, D.B. Dixon, K.C. Marsh, B. Nguyen, T.J. Opgenorth and T.W. von Geldern, Pyrrolidine-3-carboxylic acids as endothelin antagonists. 2. Sulfona-mide-based ETA/ETB mixed antagonists, *J. Med. Chem.*, 1997, **40**, 3217–3227.

3. T. Taverne, O. Diouf, P. Depreux, J.H. Poupaert, D. Lesieur, B. Guardiola-Lemaitre, P. Renard, M.C. Rettori, D.H. Caignard and B. Pfeiffer, Novel benzothiazolin-2-one and benzoxazin-3-one arylpiperazine derivatives with mixed 5HT1A/D2 affinity as potential atypical antipsychotics, *J. Med. Chem.*, 1998, **41**, 2010–2018.

4. D.R. Flower, Modelling G-protein-coupled receptors for drug design, *Biochim. Biophys. Acta*, 1999, **1422**, 207–234.

5. F. Jacob, Evolution and tinkering, *Science*, 1977, **196**, 1161–1166.

6. J. Bockaert and J.B. Pin, Molecular tinkering of G protein-coupled receptors: an evolutionary success, *EMBO J.*, 1999, **18**, 1723–1729.

7. D.C. Teller, T. Okada, C.A. Behnke, K. Palczewski and R.E. Stenkamp, Advances in determination of a high-resolution three-dimensional structure of rhodopsin, a model of G protein-coupled receptors (GPCRs), *Biochemistry*, 2001, **40**, 7761–7772.

8. R.J. Lefkowitz, The superfamily of heptahelical receptors, *Nature Cell Biol.*, 2000, **2**, E133–E136.

9. R.A. Hall, R.T. Premont and R.J. Lefkowitz, Heptahelical receptor signaling: beyond the G protein paradigm, *J. Cell Biol.*, 1999, **145**, 927–932.

10. J.J. Marinissen and J.S. Gutkind, G-protein-coupled receptors and signaling networks: emerging paradigms, *Trends Pharmacol. Sci.*, 2001, **22**, 368–376.

11. C. Discala, X. Benigni, E. Barillot and G. Vaysseix, DBcat: a catalog of 500 biological databases, *Nucleic Acids Res.*, 2000, **28**, 8–9.

12. T. Doerks, A. Barioch and P. Bork, Protein annotation: detective work for function prediction, *Trends Genetics*, 1998, **14**, 248–250.

13. D.J. Lipman and W.R. Pearson, Rapid and sensitive protein similarity searches, *Science*, 1985, **227**, 1435–1441.

14. S.F. Altschul, T.L. Madden, A.A. Schaffer, J. Zhang, Z. Zhang, W. Miller and D.J. Lipman, Gapped BLAST and PSI-BLAST: a new generation of protein database search programs, *Nucleic Acids Res.*, 1997, **25**, 3389–3402.

15. K. Hoffmann, *Protein classification and functional assignment*, in *Trends Guide to Bioinformatics*, Elsevier, 1998, pp.18–21.

16. T.K. Attwood, The quest to deduce protein function from sequence: the role of pattern databases, *Int. J. Biochem. Cell Biol.*, 2000, **32**, 139–155.

17. K. Hofmann, P. Bucher, L. Falquet and A. Bairoch, The PROSITE database, its status in 1999, *Nucleic Acids Res.*, 1999, **27**, 215–219.

18. T.K. Attwood, M.D. Croning, D.R. Flower, A.P. Lewis, J.E. Mabey, P. Scordis, J.N. Selley and W. Wright, PRINTS-S: the database formerly known as PRINTS, *Nucleic Acids Res.*, 2000, **28**, 225–227.

19. J.G. Henikoff, E.A. Greene, S. Pietrokovski and S. Henikoff, Increased coverage of protein families with the blocks database servers, *Nucleic Acids Res.*, 2000, **28** (1), 228–230.

20. A. Bateman, E. Birney, R. Durbin, S.R. Eddy, K.L. Howe and E.L. Sonnhammer, The Pfam protein families database, *Nucleic Acids Res.*, 2000, **28**, 263–266.

21. R. Apweiler, T.K. Attwood, A. Bairoch, A.Bateman, E. Birney, M. Biswas, P. Bucher, L. Cerutti, F. Corpet, M.D.R. Croning, R. Durbin, L. Falquet, W. Fleischmann, J. Gouzy, H. Hermjakob, N. Hulo, I. Jonassen, D. Kahn, A. Kanapin, Y. Karavidopoulou, R. Lopez, B. Marx, N.J. Mulder, T.M. Oinn, T.M. Pagni, F.Servant,

C.J.A. Sigrist and E. Zdobnov, The InterPro database, an integrated documentation resource for protein families, domains and functional sites, *Nucleic Acids Res.*, 2001, **29** (1), 37–40.

22. T.K. Attwood and J.B.C. Findlay, Fingerprinting G-protein-coupled receptors, *Protein Eng.*, 1994, **7**, 195–203.
23. T.K. Attwood, A compendium of specific motifs for diagnosing GPCR subtypes, *Trends Pharmacol. Sci.*, 2001, **22**, 162–165.
24. K. Palczewski, T. Kumasaka, T. Hori, C.A. Behnke, H. Motoshima, B.A. Fox, I. Le Trong, D.C. Teller, T. Okada, R.E. Stenkamp, M. Yamamoto and M. Miyano, Crystal structure of rhodopsin: A G protein-coupled receptor, *Science*, 2000, **289**, 739–745.
25. R. Henderson, J.M. Baldwin, T.A. Ceska, F. Zemlin, E. Beckmann and K.H. Downing, Model for the structure of bacteriorhodopsin based on high-resolution electron cryo-microscopy, *J. Mol. Biol.*, 1990, **213**, 899–929.
26. E. Pebay-Peyroula, G. Rummel, J.P. Rosenbusch and E.M. Landau, X-ray structure of bacteriorhodopsin at 2.5 angstroms from microcrystals grown in lipidic cubic phases, *Science*, 1997, **277**, 1676–1681.
27. F. Horn, R. Bywater, G. Krause, W. Kuipers, L. Oliveira, A.C. Paiva, C. Sander and G. Vriend, The interaction of class B G protein-coupled receptors with their hormones, *Receptors Channels*, 1998, **5**, 305–314.
28. F. Horn, E.M. van der Wenden, L. Oliveira, A.P IJzerman and G. Vriend, Receptors coupling to G proteins: is there a signal behind the sequence?, *Proteins*, 2000, **41**, 448–459.
29. J.M. Stadel, S. Wilson and D.J. Bergsma, Orphan G protein-coupled receptors: a neglected opportunity for pioneer drug discovery, *Trends Pharmacol. Sci.*, 1997, **18**, 430–437.
30. C. Southan, A genomic perspective on human proteases as drug targets, *Drug Discov. Today*, 2001, **6**, 681–688.
31. T.K. Attwood, M.D. Croning and A. Gaulton, Deriving structural and functional insights from a ligand-based hierarchical classification of G protein-coupled receptors, *Protein Eng.*, 2002, **15**, 7–12.

Virtual Screening of Virtual Libraries – an Efficient Strategy for Lead Generation

Darren V. S. Green

GlaxoSmithKline, MEDICINES RESEARCH CENTRE, GUNNELS
WOOD ROAD, STEVENAGE, HERTS SG1 2NY, UK

1 What and Why?

A Virtual Library may be defined as a set of chemical structures that theoretically could be made from defined reactions and starting materials. Virtual Screening[1] is the *in silico* evaluation of chemical structures against a model of biochemical efficacy – this could be by docking of structures to the crystal structure of an enyzme, a fit to a 3D pharmacophore model, or a prediction from a QSAR equation.

Currently, given a large enough budget, it is possible to purchase more than 2 million compounds from suppliers, and to screen these against a target of interest. Why then is Virtual Screening such an important area of research? The answer to this question is illustrated by Figure 1. There are $\sim 10^7$ compounds registered in *Chemical Abstracts*, around 10^4 compounds in the World Drug Index,[2] and, as already mentioned, $\sim 10^6$ compounds available for purchase. In contrast, estimates of the number of 'drug like' molecules which could be synthesised vary wildly, but average around 10^{30}.[3] To put this figure in some kind of context, the number of seconds since Big Bang started all of this is 10^{17}. Therefore, it is not possible to make and test everything: we must sample this huge chemical space. The success rate of random high throughput screening in identifying good start points is generally quoted as 1 compound per 10^5 compounds screened for tractable targets. This then is a base figure for the expected success rate if we sample the 10^{30} structures at random. By simple extrapolation, to be sure of finding compound series of interest for the 1000s of targets in the genome would necessitate synthesis on a scale beyond comprehension.

Therefore, we must do something other than randomly sample this virtual collection of compounds. The VSVL approach to this problem leverages two techniques which in themselves have tried to improve the efficiency of drug discovery – computer aided drug design and combinatorial chemistry. One

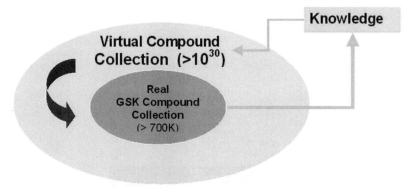

Figure 1 *Any corporate compound collection is a tiny subset of the world of chemical structures. VSVL enables the efficient exploration of this world*

major problem with computational drug discovery has been to build in synthetic chemistry tractability to molecules designed to fit receptor sites. One major problem of combinatorial chemistry has been the lack of success in making large numbers of compounds with no biological target in mind. VSVL attempts to remove both deficiencies – combinatorial libraries are by definition chemically tractable, and computer aided design should help direct the synthesis of biologically relevant molecules.

2 The Design Process – Monomer Selection and Library Enumeration

Figure 2 illustrates the process which is promoted at Stevenage *via* our ADEPT design tools.[4] Starting from databases of monomers, chemical reactions and perhaps some target knowledge, lists of monomers are produced, then refined to leave a set of possible monomers. These are then normally enumerated to the corresponding products. Product *versus* monomer based design has, in the past, been the subject of considerable debate. Monomer based design has the advantage of being quick and easy, and if your design criteria are additive – for example molecular weight and algorithms such as ClogP can be considered additive – then this can be very effective. In contrast, product based design is harder due to the numbers of products to be considered. However, product based design has been shown to be superior for 'diversity' libraries,[5] and current methods for computing 3D properties such as shape and docking scores in general require a product structure. It is proper at this point to consider a practical note which for once works in favour of a computational approach. It is possible to construct huge virtual libraries for 3- and 4-component combinatorial chemistry, from lists of apparently available monomers. In practice we find that a combination of availability and chemical reactivity so reduces the probability of these monomers actually making it into a final library that great care is taken to include starting materials which are desirable from a molecular property, ease of availability and

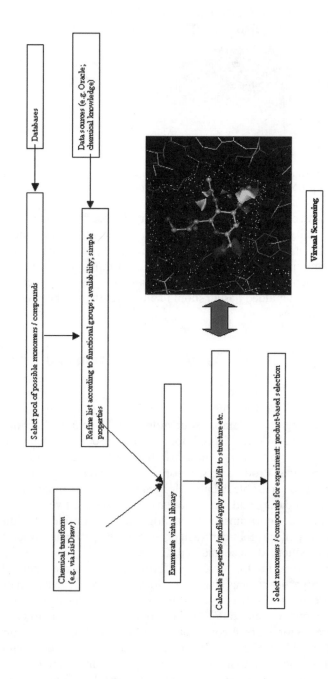

Figure 2 *The library design process*

chemical reactivity considerations. The resultant loss of potential monomers greatly reduces the size of the typical virtual library.

A method that can be used to propose desirable monomers which would justify some effort to acquire is RECAP (Retrosynthetic Compositional Analysis Procedure, Figure 3).[6] This method provides a disassemblage of known molecules at acceptable synthetic points, to generate better monomers for reassembly into libraries. It is particularly useful for biological 'systems' such as kinases, to identify so-called privileged structures.

With monomer sets and synthetic route identified, the virtual library products must be enumerated. ADEPT uses a reaction-based enumeration, using the Daylight SMIRKS language.[7] This is true *in silico* chemistry – molecule structures and chemical reactions are mixed together to create new products, if the starting materials possess functional groups that will react together. The advantages of reaction-based enumeration include the automated creation of a corporate reaction database, by the capture of reactions drawn into ADEPT, the reuse of reaction protocols to provide user-friendly tools, and the fact that ADEPT encourages the use of sequential, individual reaction steps which mirror the physical synthesis process, thus in theory providing a means to drive synthesis automation.

3 The Design Process – Virtual Screening

Once the virtual library products are enumerated, they can be profiled in a bewildering variety of methods. The simplest are property rules such as the

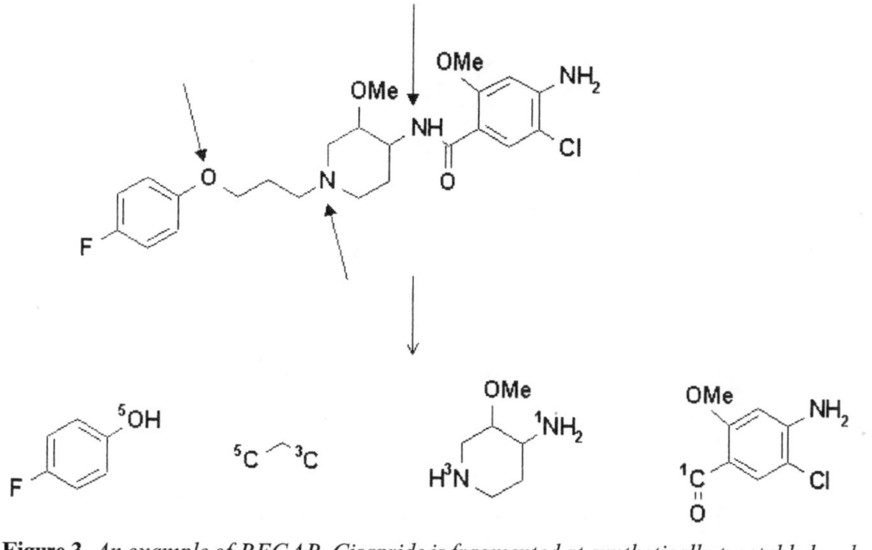

Figure 3 *An example of RECAP. Cisapride is fragmented at synthetically tractable bonds. Application to a large set of structures with related biological activity yields commonly occurring, or privileged, fragments. These fragments may then be converted into possible monomers by database searches*

Lipinski rule of 5.[8] More complicated procedures are often used, because there can be a great deal of target information available, for example crystal structures of proteins or a natural ligand. Here, we should consider the objectives of Virtual Screening, and how they differ from more familiar medicinal chemistry lead optimisation objectives.

Figure 4 illustrates the objective of computational models for medicinal chemistry projects – the construction of a mathematical description of molecules with their receptor that can predict whether molecule $n + 1$ in the series will be more active than molecule n. Therefore, the desired model must not only discriminate the black (poorly active) molecules from the grey (highly active) molecules, but also predict 'good' grey molecules from 'better' greys. Figure 5 looks at this problem from a virtual screening perspective. Obviously, it would be most effective to correctly and reliably predict affinity. However, because we are aiming to make a sufficiently large number of compounds (100s to 1000s), it is enough to discriminate grey from black, because the biological assay will discriminate 'good' grey molecules from 'better' grey molecules. Of most relevance to Virtual Screening is the need to maximise the number of compounds in the area marked 'A' (predicted and observed to be 'active') whilst minimising those compounds in 'B' (predicted inactive but actually active) and 'C' (predicted active but actually inactive). It is this balance that determines the efficiency, and hence the utility, of the VSVL processes. The objective is to do sufficiently better than random to make the whole process of generating a lead more efficient. If random screening produces a 1 in 10^5 hit rate, then it may be sufficient for VSVL to produce a $1:10^3$ hit rate, *i.e.* an enrichment of 100-fold over random. In other words, the selection process can afford to be wrong 99% of the time and still be

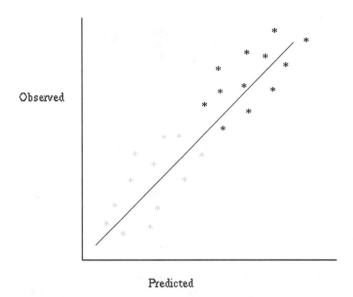

Figure 4 *The typical requirement from a lead optimisation QSAR model*

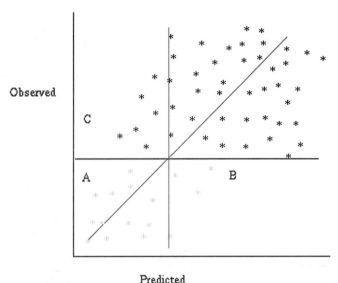

Predicted

Figure 5 *The requirement of a virtual screen is to maximise the number of molecules in region A whilst minimising the number of molecules in regions B and C*

very effective! Moving this argument from the abstract to reality, should a series of compounds have two potential binding modes which a docking algorithm cannot distinguish accurately, something that would severely hamper a medicinal chemistry program, it is acceptable for the VSVL approach to only find examples of one binding mode. The overriding objective is to discover the series in the first place, thus enabling the exploitation of the structural type, and the identification of other binding mode(s) through the panoply of methods available for lead optimisation.

Methods for virtual screening are plentiful,[9-12] the most common employed at Stevenage being 3D pharmacophore construction alongside 3D database searching. This method has been found to give very good results for focussed screening strategies.[13-15]

4 The Design Process – Library Design

In general, it can be useful to separate library design from the screening process outlined above. This is because there are many criteria that one may choose to use in library design, and these will vary from project to project. All of the algorithms described below are independent of the method used to select the desired products. In this context, then, library design is essentially monomer selection, but at this stage it is also possible to probe, automation restrictions permitting, the library configuration. For example, for an A + B + C library, what should the relative proportions of the monomer sets be? With ever-increasing flexibility of synthetic chemistry automation, it can also be instructive to use these methods to probe the usefulness of particular synthetic kit for the task – for example, can a true combinatorial solution be devised, allowing the use of split-mix type synthesis, or does the library require more flexible synthetic procedures to ensure key molecules are made outside of a true combinatorial solution? There are a variety of design tools available,[16] ranging from simple to complex algorithms. At Stevenage we nurture and encourage use of a selection of algorithms, chosen to match the problem in hand. The very simplest method is MFA (Monomer Frequency Analysis).[17] Here, those products which score well in the virtual screen are decomposed into their monomers. The number of times each monomer appears in a desired product is counted. It is then assumed that those monomers which appear most often will, when combined, produce an effective library containing many of the desired products. For single selection criteria, such as picking compounds which fit a pharmacophore, MFA can often be very effective. However, in many cases it is necessary to use a more sophisticated algorithm. PLUMS[18] is an algorithm designed to allow the optimisation of the effectiveness and efficiency of a library. Effectiveness is defined as the number of desired compounds made, divided by the total number of desired compounds in the set. Efficiency is the number of desired compounds made divided by the size of the real library. For example, 126 compounds are selected by a pharmacophore search on a virtual library of 1000. A library of 160 compounds is designed, which will succeed in making 101 of the 126 desired products. The efficiency of this library is 0.6 (101/160) and the effectiveness 0.8 (101/126). PLUMS aims to increase the efficiency of the library whilst keeping the effectiveness high. It does this by sequentially removing the worst monomer from the set – for example if a monomer appears in none of the desired products, discarding it would not change the effectiveness of the library, but would increase the efficiency. Monomers are generally removed until a user-defined library size is achieved. The results can be plotted to aid interpretation, and there is often a clear idea of what the best physical library to make will be (Figure 6). PLUMS typically takes a few minutes to run.

When more complex criteria are used, particularly the setting of multiple objectives, even more advanced methods are required. For example, the objective might be to design a library with a high number of compounds matching a pharmacophore, with a distribution of molecular weight which resembles the WDI, and which uses the cheapest monomers possible. In this case a program such as SELECT[19] can be employed. For most combinatorial libraries there are too many solutions to evaluate in order to find the best one, and so a stochastic algorithm is used. SELECT uses a Genetic Algorithm (GA) to evolve libraries which can be evaluated against the design criteria. Good solutions are kept and allowed to breed, whilst the poorest solutions are removed from the population. In this way, libraries with quite demanding design criteria can be generated. This type of tool is often used by specialists, and can take several hours to complete, depending on the size of the virtual library and the design criteria.

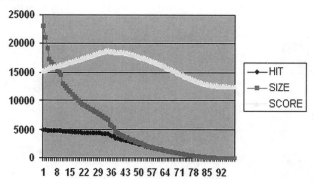

Figure 6 *Example of PLUMS output. The score is an addition of the effectiveness and efficiency of the library. The peak in the score is often indicative of the best libraries to make*

5 A Real Example

This example has been chosen, not to be the best example of VSVL, but rather an as illustration of the design processes, and the integration of many of the tools described above.

The objective of the project was to discover novel, chemically tractable series of inhibitors of an enzyme, for which several chemical series were known, but the crystal structure of the enzyme was not. A 3D pharmacophore was constructed from known chemical series, using the Catalyst.[20] The pharmacophore consists of five features (Figure 7). Concurrently, the chemical structures of known ligands for this family of enzymes were collated from the literature, resulting in more than 3000 unique compounds. RECAP was applied to this set of structures, and 1323 available, or accessible monomers were identified that encoded common chemical functionality found in inhibitors of this protein family. An ADEPT reaction scheme that could be applied to a subset of these monomers was identified and used to enumerate a virtual library of 31K products. Physicochemical properties were then computed for the products, again with ADEPT, and, by the use of Lipinski-style filters, the set was reduced to 10K desirable structures. This number was impractical for the available synthesis automation, and therefore a 3D database was constructed for the 10K products. Subsequent searching of this database using the pharmacophore selected 4900 products that fit the pharmacophore. These, then, were the products that should be synthesised. However, they did not comprise a combinatorial solution, which the synthetic equipment required. PLUMS was applied to identify the best combinatorial solution which would synthesis as many of the desired products without making a very large library. To make all the desired products, a library of 22,932 would be needed. PLUMS identified a solution whereby 3542 (72%) of the 4907 desired products could be made with a library of just 5390 products. A salutary lesson can now be imparted. With the final design in hand, it was found that many of the monomers selected were no longer available in sufficient

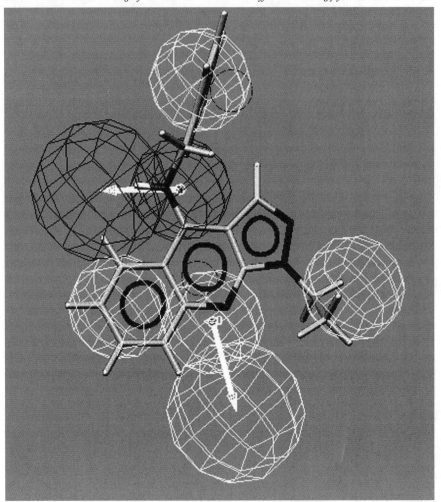

Figure 7 *The 3D pharmacophore used in the example. The triangle of light spheres indicate hydrophobic regions, the overlapping light spheres at the bottom a hydrogen bond acceptor vector, and the overlapping dark spheres a hydrogen bond donor vector*

quantity, or would not react. The actual library synthesised was therefore only 2700 compounds, many of the most attractive structures being absent. It is now common at Stevenage to build virtual libraries using monomers of more secure supply, and which are known – or, from an expert perspective are almost certain – to work in the required reaction scheme. The bottom line for every VSVL experiment is whether a series of active molecules is identified. In this case some moderately potent, novel compounds were identified (Figure 8), the best being 1 μM. At the same time, the pharmacophore was used to search the GlaxWellcome compound collection. 2000 compounds were selected for screening, using similar filters to those applied to the VSVL compounds. From these, 32 active compounds were found, in multiple chemical series, many being more potent

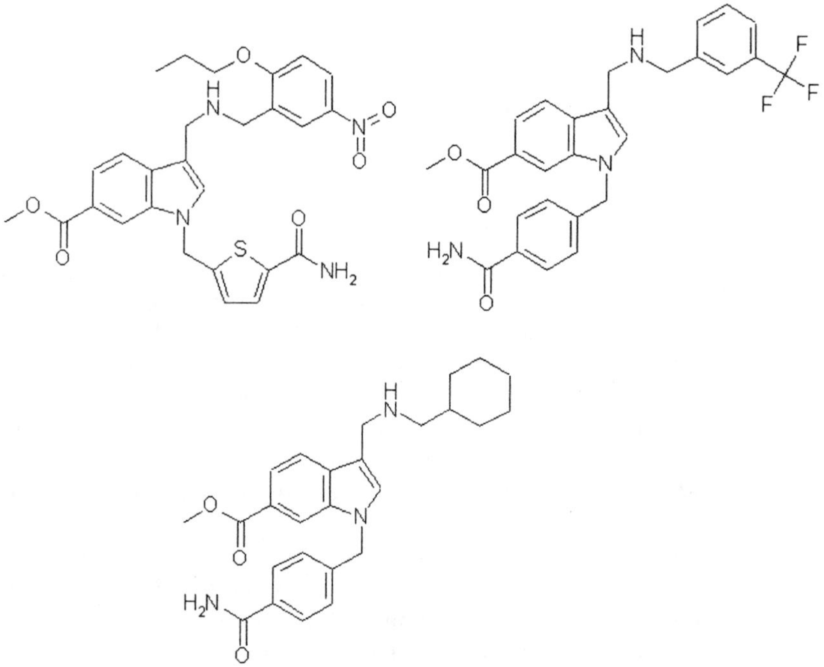

Figure 8 *The lead molecules identified using VSVL*

than the VSVL series. In our experience, this is a general result. VSVL can work, but is often less productive than selecting from a compound collection. However, VSVL does have the major advantage of adding new chemotypes outside of the existing collection, and in the final part of the paper some current efforts to address the deficiencies of the present technologies are described.

6 The Near Future – Addressing Deficiencies in the Current Technology

In our example above, the effort focused around a particular reaction scheme. It would be highly beneficial to study many schemes. One of the practical limiting factors is the time taken to evaluate multiple routes. For very large libraries, the time for enumeration and property calculation can become prohibitive. For a solution to this problem, algorithms originally developed for Patent searching can be employed. The majority of chemical patents are expressed in a Markush scheme (Figure 9). This will immediately be recognisable as how most chemists think of their combinatorial libraries. In fact, combinatorial libraries are a subset of the Markush patent requirements, that is, it is a simpler problem. CLUM-BER[21–23] builds a data structure from which all products of a library may be constructed, but which stores common fragments and monomers only once (Figure 10). This may then be used not only to enumerate product structures, but calculate additive properties (such as Lipinski-type properties), even cluster the

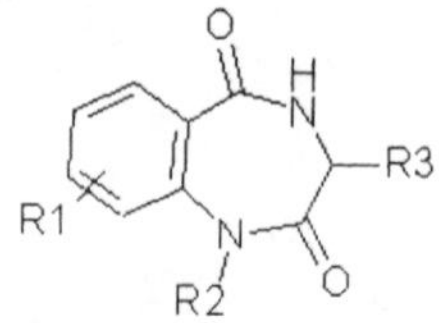

Figure 9 *A typical Markush representation*

products or compare the structural overlap between libraries. In a collaboration between GW and Barnard Chemical Information, a version of this program which can take our preferred reaction-based input and produce a Markush model has been developed. As an example for how efficient this can be, a 1 million member (100 × 100 × 100) benzodiazepine library (Figure 10) can be fully enumerated in 26 seconds[24] (38,755 products/sec), the Lipinski descriptors calculated in 96 secs (10,460 products/sec), and 2D fingerprints – needed for similarity searching or clustering on structure – in 363 seconds (2754 products/sec).

The construction of 3D databases can also be a bottleneck. To some extent this can be reduced by filtering the 2D structures, so that only compounds that have acceptable properties ever make it into 3D. However, this is still an area of active research. It may be that the CLUMBER technology may be useful in the future. One way to save effort is to keep the databases once built, and to monitor which products, built from which monomers, are already built. Then one would only need to build the new products – this is like a poor man's computer cache.

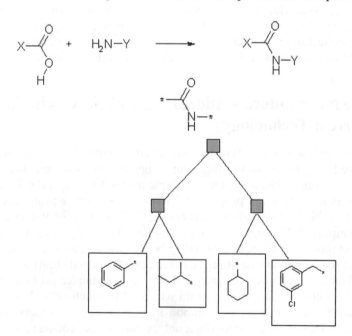

Figure 10 *The CLUMBER data model for a simple amide library*

At GW we collaborated with Silicon Graphics Inc. (SGI) and Molecular Simulations Inc. (MSI) to apply distributed computing to this problem with the catCrunch project. Catalyst 3D databases of 5.6 million virtual compounds were constructed on a 128 processor machine. This took 11,000 processor hours, with the databases occupying 90 Gb of disk space. However, it is possible to search these databases with a 3D pharmacophore at the rate of 20 minutes/1 million products. This indicates the level of throughput one might achieve by harnessing commodity computing, such as LINUX farms, or the screen-saver technology created by the SETI project.[24] Indeed, it is now possible for people to engage in a real-life virtual screening experiment organised by the cancer research charities.[25] This project is using a docking algorithm to known protein structures, and follows in the steps of a previous 'Crunch' project between Protherics and SGI.[26]

The brute force approach of these Crunch projects is likely to be succeeded by algorithmic solutions. This is because the chemistry is combinatorial, and the brute force only parallel. For example, doubling each of the monomer sets in a three-component library increases the number of products eight-fold, and doubling the number of computer processors available does not provide a solution. Docking algorithms are in the lead in this respect, where groups are trying to convert the combinatorial problem into a linear one, so that increase in library throughput maps perfectly onto increase in computer horsepower. Typically these algorithms involve docking monomers or core parts of the molecule then being smart about how these might join to other monomers, within the constraints of the protein environment (Figure 11).[27–29]

Even supposing that the virtual screen performs as well as we would like, there still remains the practical problem of reducing the set of desired products to an efficient synthetic process. With state of the art automation, it is not strictly necessary to have truly combinatorial solutions, but even so it is desirable to be as near combinatorial as possible to reduce reagent numbers and costs. As has been discussed above, this is a multiobjective optimisation problem, best tackled with a stochastic algorithm working on product properties. SELECT, and similar programs,[30,31] use a weighted sum approach to scoring a library. For example, should one wish to optimise the diversity of a library, whilst restraining the molecular weight profile to mimic that of the WDI and keeping the reagent costs low, each proposed library would be scored by summing the individual contributions from these objectives, each weighted by a user-defined amount. There are several problems with this approach. The first is that, without experimentation, there is no sensible way of choosing the weights, and thus a lot of time must be expended to observe the effect of the weights on the monomers eventually chosen. In addition, the use of many objectives may result in a uniformly average solution across all the objectives, when actually there may be an optimum solution for all but one of them. Given this information, the chemist might well decide that this particular objective could be relaxed, given the excellent performance against the others. With a weighted sum scoring function, it is not possible to provide this information. For a solution to this problem, we have applied the principles of Pareto Optimality and implemented a MultiObjective Genetic Algorithm (MOGA).[32–35] These methods are used for decision support

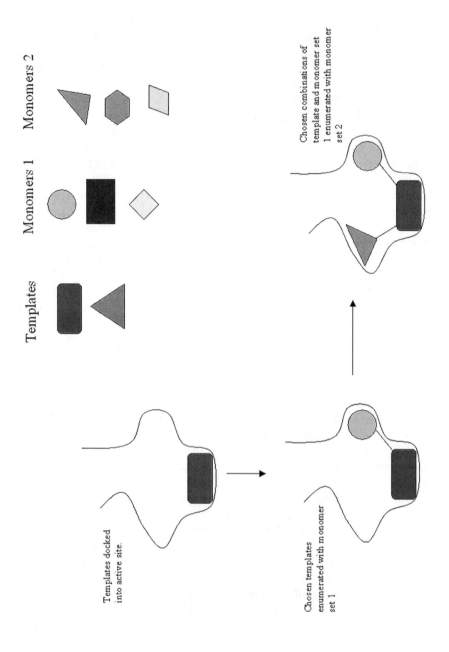

Figure 11 *Illustration of a docking procedure designed to avoid the docking of every combinatorial product*

in engineering control systems. The principles of pareto optimality are illustrated in two dimensions for a simple problem – how to find solutions which simultaneously optimise the values of x and y? The stars represent potential solutions, and the purpose of the algorithm to find the Pareto Frontier (the curved line in Figure 12). This is a set of *non-dominated* solutions. A non-dominated solution is when a line can be drawn from the star to the x and y axes, and no other stars fall into the box enclosed by these lines and the axes. In Figure 12, each star is annotated by the number of solutions that dominate that star. The non-dominated solutions are marked with a 0, and are close to the Pareto Frontier. Figure 13 brings the theory into practice. This is the result of running the MOGA to choose a 10K member library from a 100K virtual library, with the objective to optimise diversity whilst restraining the molecular weight profile to that of the WDI. Several observations can be made from the range of solutions identified. The first is that there are many similar solutions along each part of the Pareto Frontier. Traditional weighted sum approaches will return a solution, without the information that there are many neighbouring solutions that may be deemed superior upon other criteria, for example a near-by solution may contain more attractive monomers to the 'chemist's eye' than the solution returned by the algorithm. Secondly, the two objectives are in competition, and that it is not possible to obtain a solution with optimal criteria for both objectives. The final observation is that the algorithm does a good job in providing a range of solutions across the Pareto Frontier, and it happens to do this in the time taken for a traditional algorithm to produce it's 'best' solution. The challenge now is to provide ways to visualise and interrogate all the data that is produced.

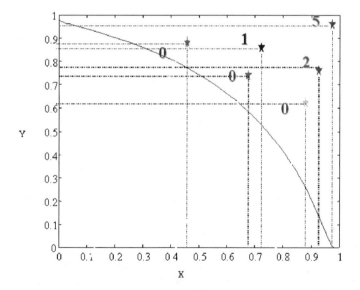

Figure 12 *An illustration of Pareto ranking for possible solutions to a problem where two of the objectives are to optimise the values of X and Y. The curved line, the* Pareto frontier, *shows the desired set of solutions*

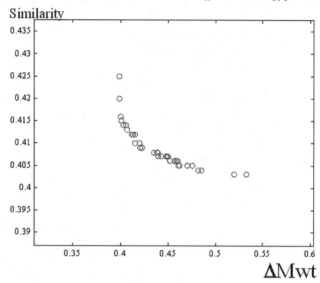

Figure 13 *The set of libraries designed using a MOGA. Each point is a different library, with the solutions spread over the range of two objectives – internal similarity of the library, and the difference in the molecular weight profile of the library compared to the WDI. The two objectives can be seen to be in competition*

7 Conclusions

VSVL is a proven way of finding novel leads from large virtual libraries. It is not yet the 'efficient strategy for lead generation' promised in the title of the paper. This is due to several reasons, some scientific, some technical. The algorithms and scoring functions required to predict interactions between ligand and protein are under constant improvement. The less rigorous requirements for VSVL over lead optimisation applications has enabled much progress in the area, with the result that there are several validated schemes for ligand docking, and there is a growing literature documenting the success of 3D pharmacophore methods. Library design algorithms continue to improve, with the move towards MOGAs an illustration of how sophisticated and effective these might become. Smarter representation of the combinatorial libraries themselves is enabling huge leaps in efficiency.

In general, however, these tools remain quite daunting for the majority of scientists working at the bench. The routine use of these methods is not simply a matter of education and training. Integration of these methods into accessible and robust end user software, without dilution of the science, must be a priority. In this respect the software we have can be compared to the fledgling synthesis automation of the 1990s. The future should see a shift of emphasis from synthesis hardware to design software. The author, for one, cannot wait.

Acknowledgements

This paper owes it's existence to the work of many colleagues and collaborators: Andrew Leach, Mike Hann, Giampa Bravi, Nick Bailey, Xiao Qing Lewell, Duncan Judd, Steve Watson, Graeme Robertson and Francis Atkinson from GSK; Al Maynard from MSI; Val Gillet, Peter Willett, Peter Fleming and Illy Khatib from the University of Sheffield; John Bradshaw and Jack Delaney of Daylight CIS; John Barnard and Geoff Downs of Barnard Chemical Information.

References

1. *Virtual Screening for Bioactive Molecules*, ed. H.-J. Boehm and G. Schneider, Wiley-VCH, Weinheim, 2000.
2. *The World Drug Index* (WDI) is available from Derwent Information, 14 Great Queen St., London WC2B 5DF, UK.
3. Y. C. Martin, *Perspect. Drug Disc. Des.*, 1997, **7**, 159.
4. A.R. Leach, J. Bradshaw, D.V.S. Green, M.M. Hann and J.J. Delany III, *J. Chem. Inf. Comput. Sci.*, 1999, **39**, 1161.
5. V. J. Gillet, P. Willett and J. Bradshaw, *J. Chem. Inf. Comput. Sci.*, 1997, **37**, 731.
6. X. Q. Lewell, D. Judd, S. Watson and M. Hann, *J. Chem. Inf. Comput. Sci.*, 1998, **38**, 511.
7. *Daylight Theory Manual*, Chapter 7, Daylight Chemical Information Systems, Santa Fe, and http://www.daylight.com/dayhtml/doc/theory/theory.rxn.html
8. C. A. Lipinski, F. Lombardo, B. W. Dominy and P. J. Feeney, *Adv. Drug. Deliv. Rev.*, 1997, **23**, 3.
9. W. P. Walters, M. T. Stahl and M. A. Murcko, *Drug Disc. Today*, 1998, **3**, 160.
10. T. J. Marrone, B. A. Luty and P. W. Rose, *Perspect. Drug. Disc. Des.*, 2000, **20**, 209.
11. C. A. Baxter, C. W. Murray, B. Waszkowycz, J. Li, R. A. Sykes, R. G. Bone, T. D. Perkins, W. Wylie, *J. Chem. Inf. Comput. Sci.*, 2000, **40**, 254.
12. C. Bissantz, G. Folkers and D. Rognan, *J. Med. Chem.*, 2000, **43**, 4759.
13. T. Langer and R. D. Hoffmann, *Curr. Pharm. Des.*, 2001, **7**, 509.
14. J. S. Mason, A. C. Good and E. J. Martin, *Curr. Pharm. Des.*, 2001, **7**, 567.
15. D. P. Marriott, I. G. Dougall, P. Meghani, Y.-J. Liu and D. R. Flower, *J. Med. Chem.*, 1999, **42**, 3210.
16. D. H. Drewry and S. S. Young, *Chemomet. Int. Lab. Sys.*, 1999, **48**, 1.
17. W. Zheng, S. J. Cho and A. Tropsha, *J. Chem. Inf. Comput. Sci.*, 1998, **38**, 251.
18. G. Bravi, D. V. S. Green, M. M. Hann and A. R. Leach, *J. Chem. Inf. Comput. Sci.*, 2000, **40**, 1441.
19. V. J. Gillet, P. Willet, J. Bradshaw and D. V. S. Green, *J. Chem. Inf. Comput. Sci.*, 1999, **39**, 169.
20. Catalyst, available from Molecular Simulations Inc., 9685 Scranton Road, San Diego, CA 92121, USA.
21. J. M. Barnard and G. M. Downs, *Perspect. Drug. Disc. Des.*, 1997, **7**, 13.
22. J. M. Barnard, G. M. Downs, A von Scholley-Pfab and R. D. Brown, *J. Mol. Graphics Mod.*, 2000, **18**, 452.
23. G. M. Downs, J. M. Barnard, *J. Chem. Inf. Comput. Sci.*, 1997, **37**, 59.
24. http://www.daylight.com/meetings/mug00/Barnard/
25. http://setiathome.ssl.berkeley.edu/

26. http://www.ud.com/projects/cancer/
27. http://www.protherics.com/crunch
28. G. Jones, P. Willett, R.C. Glen, A.R. Leach and R. Taylor, *ACS Symp. Ser.*, 1999, **719**, 271.
29. M. L. Lamb, K. W. Burdick, S. Toba, M. M. Young, A. G. Skillman, X. Zou, J. R. Arnold and I. D. Kuntz, *Proteins: Struct., Funct., Genet.*, 2001, **42**, 296.
30. F. L. Stahura, L. Xue, J. W. Godden and J. Bajorath, *J. Mol. Graph. Mod.*, 1999, **17**, 1.
31. W. Zheng, S. T. Hung, J. T. Saunders and G. L. Seibel, *Pacific Symposium on Biocomputing*, 2000, 588.
32. Cerius2 Combichem, available from Molecular Simulations Inc., 9685 Scranton Road, San Diego, CA 92121, USA.
33. V. Gillet, W. Khatib, P. Willett, P. Fleming and D. V. S. Green, *J. Chem. Inf. Comput. Sci.*, 2002, **42**, 375; Intl. Patent applied for.
34. C. M. Fonseca and P. J. Fleming, *Genetic Algorithms for Multiobjective Optimization: Formulation, Discussion and Generalization*, in *Genetic Algorithms*, Proceedings of the Fifth International Conference, San Mateo, CA, ed. S Forrest, Morgan Kaufmann, 1993.
35. C. M. Fonseca and P. J. Fleming, *An Overview of Evolutionary Algorithms in Multiobjective Optimization*, in *Evolutionary Computation*, ed. K. De Jong, The Massachusetts Institute of Technology, 1995, Vol. 3, No. 1, pp. 1.
36. W. Khatib and P. J. Fleming, *The StudGA: A Mini Revolution?*, in *Parallel Problem Solving from Nature – PPSN V*, Proceedings of 5th International Conference, ed. A. E. Eiben, T. Bäck, M. Schoenauer, H. P. Schwefel, Springer-Verlag, 1998, pp. 683.

Virtual Techniques for Lead Optimisation

Iain M. McLay

GlaxoSmithKline, MEDICINES RESEARCH CENTRE, GUNNELS
WOOD ROAD, STEVENAGE, HERTS SG1 2NY, UK

1 Introduction

In modern pharmaceutical research the task of lead finding and lead optimisation are seen as two distinct functions. Lead finding has undergone a revolution
in the past five years. Dr. Green[1] has presented many of the computational
innovations powering this revolution: focused screening; mining of high
throughput screening data; library design for high throughput chemistry and
analysis of virtual libraries. In this article virtual techniques will be considered
that could lead the way to a similar revolution in lead optimisation.

2 Background

In the drug discovery process a new lead is expected to have five key attributes:
(1) reproducible *in vitro* activity for the pure compound; (2) proven activity
through the required mechanism; (3) reasonable SAR relationships for a small set
of analogues; (4) 'chemical tractability', that is have a wide range of accessible
chemistries available for structural exploration of the compound; (5) either
acceptable 'drug-like' properties,[2] or capable of leading to analogues with acceptable drug-like properties. In addition, there is a strong opinion[3,4] that a lead
should be structurally simple so that elaboration, to provide increased potency,
is possible whilst staying within the bounds of properties considered to be
drug-like. One factor, the subject of much debate, is the *in vitro* potency required
for a lead. Opinions vary greatly from sub-micromolar to $50\,\mu$M. Leads with
high micromolar activity have certainly been modified to provide nanomolar
active compounds. A prime example of this is the development of Losartan, IC_{50}
19 nM, developed from the Takeda angiotensin II antagonist, IC_{50} $42\,\mu$M (Figure 1).[5] In truth, the criteria for acceptable lead activity varies depending on the
structure of the compound and the scope for elaboration. It often comes down to
a pragmatic statement 'will the medicinal chemist pick it up as a lead or not?'

The properties that have to be considered by the medicinal chemist during

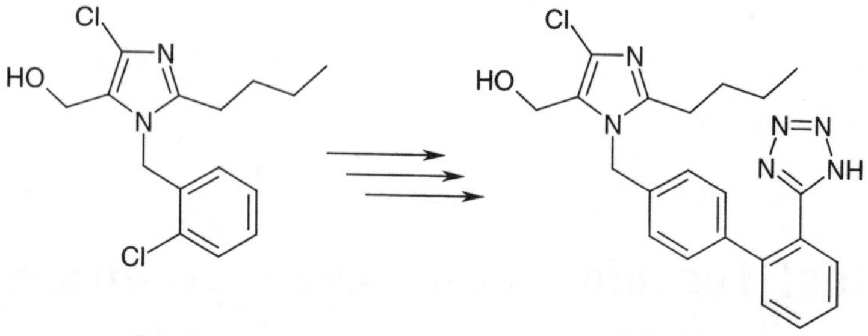

TAKEDA LEAD   LOSARTAN

Figure 1 *The development of the potent angiotensin II receptor antagonist Losartan from the weakly active Takeda lead*

lead optimisation are shown below, with arrows indicating the usual requirement to increase or decrease the property.

↑ Potency
↑ Selectivity
↑ Patent coverage

↑ Intestinal absorption
↑↓ CNS penetration
↑ Metabolic stability
↑ Solubility
↓ Protein Binding
↓ Toxicity

This is a long list of properties with the first block representing the traditional goals of potency and patents and the second block properties that have recently become known as 'developability criteria'. In modern pharmaceutical research it is critical that each one of these developability criteria be considered as early as possible in the optimisation process. It is hoped, by so doing, to reduce the proportion of failures at a late stage by building in good properties at an early one. With large Pharma groups currently spending nearly $500,000 an hour on research and development the price of failure has become very high indeed and it is no longer acceptable, even with built-in developability properties, for a lead optimisation project simply to provide a development compound and a single backup. The requirement must be for a portfolio of good compounds,[6–10] any one of which could eventually make it to market. The strategy should then be to push forward a single compound, as a pioneer, into the most expensive phases of development with the expectation of returning to the portfolio to select a replacement compound if the pioneer fails. Clearly, to achieve this increased output we have to synthesise and screen more compounds. The industry is applying several techniques in the real world to achieve these goals: (1) methods of array and library synthesis,[6,8] (2) rapid throughput developability screens[9] (Caco-2 permeability, solubility, microsomal turnover, CYP450 turnover and inhibition, cassette dosed pharmacokinetics *etc.*) and, of course, (3) rapid *in vitro* potency and selectivity screens. Computational chemistry can assist the process further by paralleling these real world activities in the virtual world. These virtual activities fall under the headings of: Virtual Focused Diversity (experimental design of arrays and libraries); Virtual Screening (potency, selectivity) and Virtual Filtering (developability criteria). Each of these will be dealt with separately below.

3 Virtual Focused Diversity for Experimental Design of Arrays and Libraries

The term 'focused diversity' is an odd one. The objective is simply to define a region of property space that is relevant to the problem in hand and then ensure that this region is spanned effectively with the compounds made. There are several design methods that have been applied by medicinal chemists in order to achieve such a goal, e.g. Fractional Factorial,[10] D-Optimal[11] and Partitioning.[12] The most important step in the process is defining the property space to be explored. There has been much discussion of diversity and how to evaluate it.[13] It is clear that pharmacophore groups in drug molecules are key to their interaction with protein targets[14–16] and, therefore, for general exploration of potency the thorough exploration of pharmacophore space is vital. Several groups have devised methods to achieve this goal.[17–19] The Gridding and Partitioning (GaP) method[19] will be discussed here in more detail. The general concept is laid out in Figure 2 for the selection of monomers to be used in an array synthesis. It may be useful to consider a real example: Figure 3 shows the synthesis of a simple array that was used in the identification of the p38 inhibitor RPR200765A.[6] The reaction involved is a straightforward amide coupling. In this work the medicinal chemists used 32 amine monomers selected to represent the common pharmacophore groups and to cover reasonable distances of those groups from the amide linkage bond. The selection, which was carried out manually (atom and bond counting) and with no quantification of pharmacophore coverage, is a very simple example of experimental design using pharmacophores. The GaP method allows such a selection to be carried out in an automated fashion and ensures the pharmacophore coverage required. The method considers a large number of possible monomers, normally selected to lie within property constraints defined by the medicinal chemist. The linkage bond to the monomer is fixed in the co-ordinate frame, which is partitioned into cubes of fixed dimensions. Each monomer is linked in turn to the template to generate a virtual molecule. Conformational analysis is performed allowing all the bonds contained in, and linked to, the monomer to vary in a systematic fashion. As the analysis continues the position of the pharmacophore groups for each conformer is identified, occupation of a partition cube is recorded and the information stored in a bit string (see Figure 2). Finally, selection of monomers is made by reviewing the molecular bit strings and selecting the smallest set of monomers that fill all the bits possible. There are many ways of modifying and biasing these selections and these are laid out in the original publication.[19] The array prepared during the

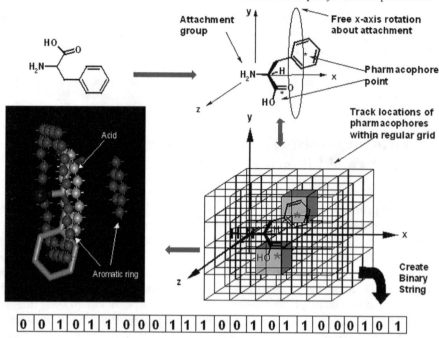

Figure 2 *Schematic description of the GaP method. Each monomer is aligned in the same co-ordinate space. Conformer generation is performed allowing rotation around all bonds. For each conformer the position of pharmacophore groups in the 3D-grid are determined and recorded in a bit string*

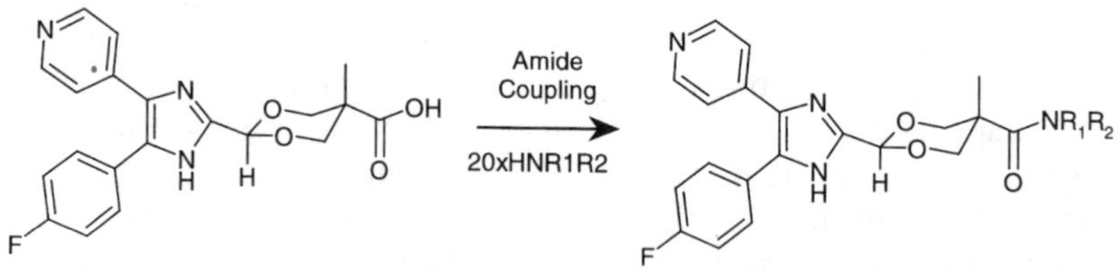

Figure 3 *Synthesis of the amide array used in the discovery of RPR200765A[6]*

discovery of RPR200765A was re-evaluated using GaP. For the study a set of 1330 suitable (MW <200, cLogP 1.5–3.5, rotatable bonds <4, pharmacophore groups <2) amines were identified in the Available Chemical Directory and processed using GaP. It was found that 150 amines were required to fill (at least once) all the partitions available to the combined set. The original selection of 32 was found to cover just 14% of the bins available to these suitable amines. Clearly the original selection left most of the property space unexplored.

4 Virtual Screening for Potency and Selectivity

These techniques may be divided into methods for which the binding site structure is known, through X-ray crystallography, and those in which it is not. These are considered below.

4.1 Site Structure Unknown

There are three familiar techniques which may be adapted for use in virtual screening: (1) pharmacophore fitting, *e.g.* using the Catalyst[24] package (Figure 4), (2) receptor surface models[20,21] (Figure 5) and (3) a statistical receptor model (CoMFA[22]/COMSIA[23]) (Figure 6). The first technique can act as a reasonably crude selection criterion with compounds showing close RMS fitting to the pharmacophore preferred over those with looser fitting. The discriminating power of the pharmacophore can be increased greatly by the addition of shape[25] or exclusion spheres.[26] However, the other two methods (2) and (3) should be capable of giving an actual prediction for the activity for the virtual compounds. Unfortunately, both suffer from the fact that they require a careful superimposition for each compound. This can be very limiting when many virtual compounds need to be evaluated. Several groups have looked at methods which may provide an alignment free method, *e.g.* the MS-Whim[27] or the Almond method.[28] Almond appears to be promising, though the method is at an early stage (Figure 7). The technique utilises a single low energy conformation for each molecule. The conformer is explored using three molecular probes, *e.g.* water probe, dry (hydrophobic) probe and amide NH. For each probe the regions of energy

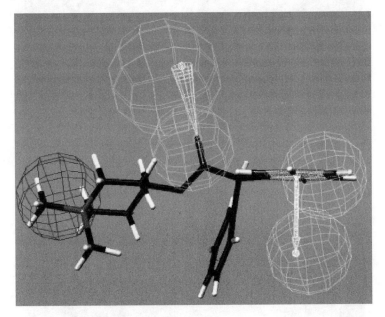

Figure 4 *A typical pharmacophore with virtual compound fitted: Positive centre (dark mesh spheres to left of picture), H-bond donor atom and extension point (light mesh spheres to centre of picture) and aromatic ring centroid with normal (light mesh sphere to right of picture)*

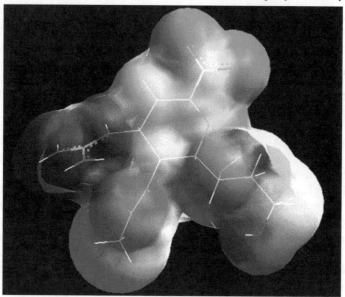

Figure 5 *A typical receptor surface model showing coding for H-bonding interactions: H-bond acceptor (light grey) and H-bond donor (dark grey)*

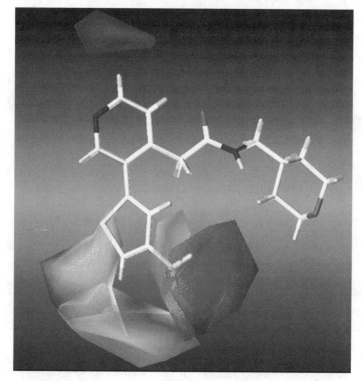

Figure 6 *A typical CoMFA model for steric interactions: favoured regions (dark grey shading to right and top of molecule) and disfavoured regions (light grey shading to right of molecule)*

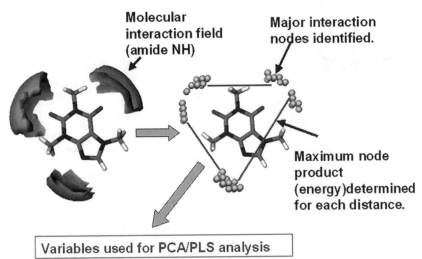

Figure 7 *Schematic description of the Almond method. Regions of favourable interactions with amide NH probe identified. These are reduced to a defined number of interaction nodes. For each combination of nodes the product of the interaction energies and the distance is recorded. The largest product for each distance is retained for multivariate analysis*

minima are identified and the distances between the GRID nodes, which comprise these regions, are then determined along with, for each pair, the product of the interaction energies. This process is performed for nodes derived from the same probe (auto) and for nodes from different probes (cross) and the maximum products for each distance stored. These distance–energy variables are then related to activity through multivariate analysis (PCA, PLS) in order to generate quantitative or qualitative models suitable for virtual screening. The method has the advantages of being an interpretable, rapid, pharmacophore based QSAR with no alignment necessary, but the disadvantage that it uses only two point pharmacophores which are derived from a single conformation. Currently we are finding the method most useful when applied to small, simple systems and using PCA rather than the PLS regression method. An example of this approach is shown in Figures 8 and 9 where a series of 7TM receptor antagonists was considered. The structures were simplified by ignoring the common portion and exploring the single variant position to which various heterocycles had been attached. The Almond analysis was carried out on the training set, PCA performed and the results displayed as a PCA scores plot with colour coding according to activity (black active, grey inactive, in Figure 9). Three regions containing active compounds were identified. A large set of possible heterocyclic replacements were screened virtually by projection of their Almond descriptors back into this PCA space. It was then possible to nominate for synthesis those virtual heterocycles found to be placed in the 'active' regions (Figure 9).

 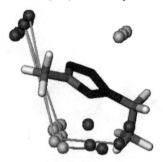

Figure 8 *Typical Almond variables for a heterocycle. Such variables were used to generate the virtual screen developed in Section 4.1. Picture to left shows a substituted tetrazole with autocorrleation descriptors derived from a GRID dry probe. Picture to the right shows same heterocycle with crosscorrleation descriptors for dry (light grey spheres) and amide NH (dark grey spheres) GRID probes*

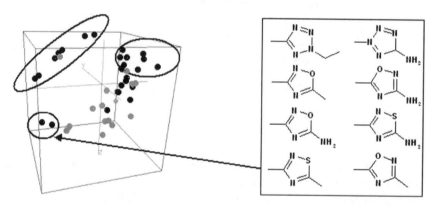

Figure 9 *Left, Principal Component Analysis plot for virtual screen described in Section 4.1, with active regions ringed. Right, heterocycles identified as projecting neatly into one of the regions found for active compounds*

4.2 Site Structure Known

It may be expected that virtual screening would be fairly straightforward when the active site structure of a target is available. What is required is a good method for docking virtual compounds and a good scoring method to rank the compounds according to the goodness of fit to the receptor. There are a huge number of docking and scoring methods available and there have recently been some good comparisons of the methods.[29] However, the common experience for this sort of work, when applied to lead optimisation, is that docking programs work extremely well, inasmuch as crystal structures can be reproduced accurately, but the relative scoring of dockings for different ligands is very poor indeed. An example of the sort of results obtained can be seen in Figure 10 for p38 MAP kinase and a series of inhibitors. It can be seen that there is no correlation of the IC_{50} with the docking score (GOLD[30]). For the moment the docking and scoring

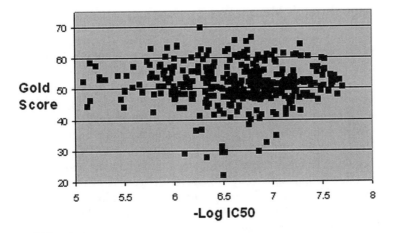

Figure 10 *Typical plot of GOLD scores against IC_{50}: no correlation can be seen. The GOLD score is in essence an estimation of the binding energy for the ligand, but with sign reversed to make large values favourable for binding*

are useful for eliminating very poor compounds ('no hopers') but cannot be used to rank the remainder. In the long term we need new accurate scoring methods. In the short term one approach is to develop a CoMFA or CoMSIA model for the target and embed this into the site for scoring. In this way a target specific scoring function is created. Such an approach was reported to be superior to all the currently available scoring methods for a set of for p38 MAP kinase inhibitors.[31]

5 Virtual Filtering for Developability Criteria

There are a great number of workers involved in the development of general models to predict developability criteria. These can be divided into two types: Level 1 models in which compounds are assessed for their 'drug-like' properties, such as application of the Lipinski Rule-of-5[2] or the Sadowski neural network classification.[32]

Level 2 models designed to predict a single property: Intestinal absorption,[33] CNS penetration,[34,35] metabolic stability,[36] solubility,[37] protein binding,[38] toxicity.[39]

For lead optimisation it is advisable to check the suitability of these models for the series under investigation. It will often be found that the Level 1 models have to be relaxed for the series and that the Level 2 models simply fail. When the Level 2 models fail it is appropriate, if sufficient measured data become available, to develop a series specific model. It is not possible here to discuss all the developability criteria, so the discussion will concentrate on three areas: passage through membranes, toxicity prediction and metabolism prediction.

5.1 Passage through Membranes

VolSurf is a new technique designed to have particular applicability to the modelling of the movement of drugs through membranes[40] and as such may have utility for prediction of absorption and penetration of the blood–brain barrier (BBB). It uses volume properties describing hydrophobic, hydrophilic, H-bonding interactions and various combinations of these properties (Figure 11). A BBB prediction model has been developed using this methodology.[41] The model was generated using data almost entirely from a single congeneric series of κ-opioid analgesics (*e.g.* Figure 12). However, it was found also to be effective for the general prediction of CNS active compounds, though rather less successful for prediction of those that were not CNS active. In order to evaluate the VolSurf method for lead optimisation a three component PCA model was constructed largely according to the published method, but with the modification that only the members of the congeneric series were included and that a single representative enantiomer was used for each racemic pair. It was found that separation of the penetrant from the non-penetrant compounds was best viewed using a plot of the first and third PC (Figure 13). A set of sedating (10 compounds) and non-sedating (10 compounds) antihistamines were then projected into this model (Figure 14). All the sedating (penetrant compounds) were seen to lie in the correct zone. However, the non-sedating were split between zones with five in the correct zone and five misclassified. Such a model could be very useful for virtual screening, but its application has to be considered carefully. When working with a large virtual library it should be possible, through this method, to identify a subset of compounds that will not pass into the CNS. There will be other non-pentrating compounds that will be misclassified, but this is acceptable if we are dealing with a large number of possibilities and limited chemical effort. Alternatively, if we are seeking CNS active compounds, the model will ensure we don't miss any potentially good compounds. However, it will also identify a large number of non-penetrating compounds as CNS active.

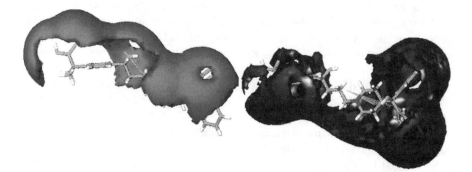

Figure 11 *Typical VolSurf properties: left favourable water interaction volumes, right favourable hydrophobic volumes for the antihistamine fexofenedine*

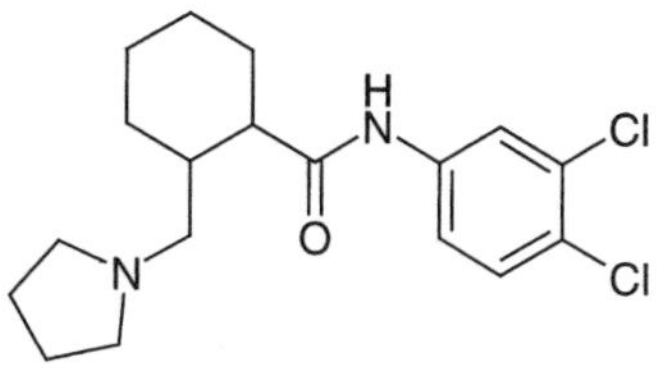

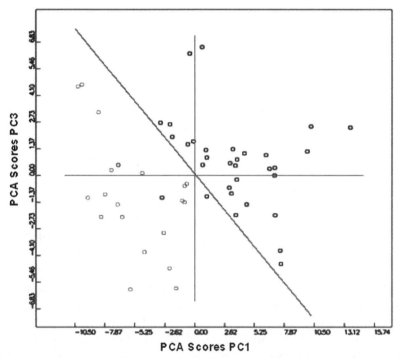

Figure 12 *Typical κ-opioid agonist used in the production of the blood–brain barrier model described in Section 5.1*

Figure 13 *PCA plot for the BBB model derived for κ-opioid agonists described in Section 5.1: CNS active compounds light grey open circles and CNS inactive compounds dark grey open circles*

5.2 Toxicity Prediction

There are several systems available for predicting well known toxicities related to substructure.[39] However, there are many possible sources of toxicity and the usual problem for lead optimisation is series related toxicity. This can be very difficult to model and is unlikely to be predicted by the general toxicity prediction systems. It is common for a lead optimisation team to be aware of a particular toxicity related to a series, but rare to generate enough whole animal toxicity data to allow production of a model. One way forward is to look for a simple *in vitro* surrogate that can generate an endpoint, related to *in vivo* toxicity, on a substantial number of compounds. An example of this is seen with the work

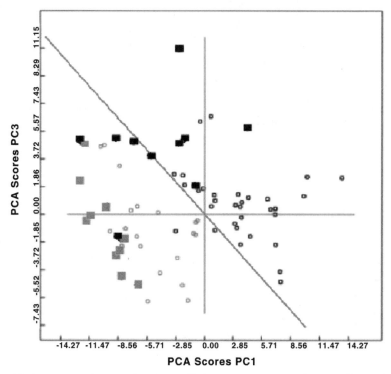

Figure 14 *Projection of a set of sedating (light grey squares) and non-sedating (black squares) anti-histamines into the PCA space defined for the κ agonist model (Figure13)*

described for the identification of RPR200765A. In this case the general class of compounds was known to have a liver toxicity problem. Initial *in vivo* rodent studies were carried out which indicated that this was most likely due to induction of CYP450 1A1. An *in vitro* CYP450 1A1 induction screen using liver hepatocytes was employed as a rapid screen for toxicity, thus allowing rapid generation of good quality toxicity related data. It is *in vitro* data of this sort that would be ideal for the development of a virtual filter. VolSurf and Almond may be used for the modelling of such data. In Figure 15 one particular cellular toxicity marker was modelled using VolSurf water, dry and amide NH probe variables and PCA. The 'toxic' compounds (light grey) were clearly seen to lie in a different portion of PCA space to the non-toxic (dark grey). Virtual filtering for toxicity was then carried out by projection of compounds into the PCA space and using the distance from the toxic region to prioritise compounds for synthesis.

5.3 Metabolism Prediction

There are many reports describing the use of QM calculations for the prediction of sites, and relative rates, of metabolism.[42] These are highly labour intensive,

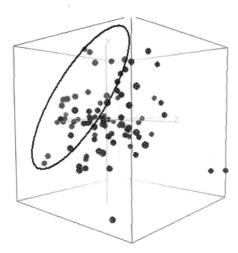

Figure 15 *An Almond virtual toxicity filter. Toxic compounds (light grey) can be seen in a distinct region separate from the non-toxic compounds (dark grey)*

complex and time-consuming calculations that are certainly not suited for use as a high throughput virtual metabolism prediction filter. An investigation was carried out to see if *in vitro* metabolism could be predicted for a series of compounds using Almond. The rational behind this is that metabolising enzymes have very simple recognition features, and that Almond's simple two point/distance–interaction energy approach may be able to simulate these features. A series of 7TM ligands, with rat microsomal turnover data, was used for this work. They were categorised into high and low turnover. Almond parameters were generated and PCA analysis was carried out. The initial three component model was simplified somewhat by using the most influential variables only and a two PCA component model produced. The two models are shown in Figure 16. An examination of the plot reveals a region in the lower left-hand quadrant that contains compounds largely free of metabolism. All other areas of the plot contain an approximately equal mixture of both extensively and poorly metabolised compounds. Clearly there are many low metabolism compounds found in the other regions. However, in this case virtual filtering would be directed at prioritising for synthesis compounds found to project into this promising left-hand quadrant.

5.4 An Example of Virtual Filtering for Library Design

The objectives of the medicinal chemists for the synthesis of the library in Figure 17[43,44] were to produce compounds passing some simple Level 1 filters (Lipinski rules, modified in the light of previous knowledge of the bioavailability of the series, and polar surface area $< 140 \text{ Å}^2$) whilst at the same time satisfying the constraints of the synthetic protocols to be used. These were: (1) a near 20×20 Library (~ 400 final compounds, but not essential to be fully combinatorial); (2) **Am** very close to twenty; (3) **Nu** number not so critical; (4) All **Am**s to be used at

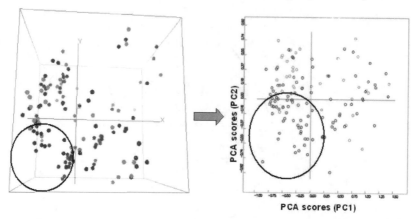

Figure 16 *A VolSurf virtual metabolism filter for a series of 7-TM antagonists. Left a PCA plot of the VolSurf descriptors (PC1-3). Right a refined model in which the variables are reduced to the most influential. A region coding for low metabolism can be seen in the lower left-hand quadrant. Virtual filtering is carried out by projection of virtual compounds into this PCA space*

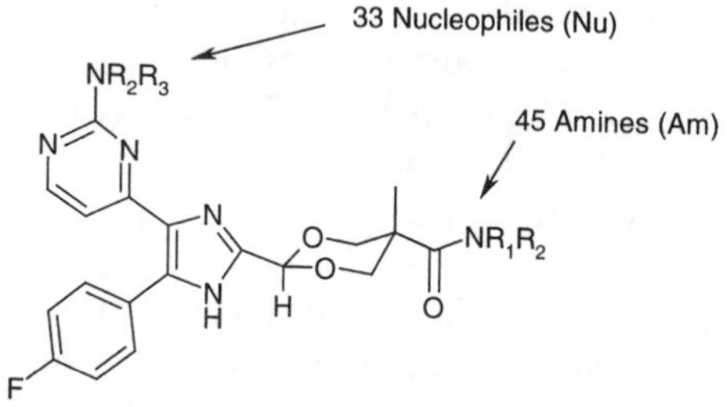

Figure 17 *Description of the Lib2 described in Section 5.4*

least 20 times. A Monte Carlo approach was used to find a solution to this problem[43] in order to satisfy all these objectives in an optimum fashion. The original virtual library contained 1485 compounds which reduced to 770 on application of the filters. The 770 compounds were processed by the Monte Carlo monomer selection to provide a library of 441 compounds using:

21 Amides → Target 20;
24 Nucleophiles → Target > 20
~21 Nucleophiles for each amide → Target 20

The success of the virtual filtering was evaluated by comparing the results of this library in a Caco-2 cell permeability assay with those for a similar library created without virtual filtering. The comparison can be seen in Figure 18. There is a clear shift to highly permeable compounds indicating success for the technique.

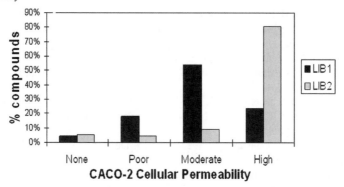

Figure 18 *Bar graphs of Caco-2 permeability data for two libraries: Lib 1 (black) synthesised without virtual filtering and Lib2 (light grey) for which virtual filtering was applied*

6 Conclusions

Virtual techniques are starting to make a big impact in lead optimisation programmes. They are sure to play a major role in the drive to produce larger numbers of high quality, robust development candidates. However, much work has yet to be done in the development, validation and application of the techniques. Pharmacophore experimental design and virtual screening are two complementary methods. One is exploring what is possible, the other quantifying it and allowing focus onto the highest quality compounds. The two techniques should both be allied to virtual developability filters. The development of these filters is the greatest challenge. General developability models will certainly play a role, but in many cases it will be necessary, when dealing with a lead optimisation series, to adapt intelligently the models or generate series specific models.

References

1. D. V. S. Green, this volume.
2. C. A. Lipinski, F. Lombardo, B. W. Dominy and P. J. Feeney, *Adv. Drug Delivery Rev.*, 1997, **23**, 3.
3. M. M. Hann, A. R. Leach and G. Harper, *J. Chem. Inf. Comput. Sci.*, 2001, **41**, 856.
4. S. J. Teague, A. M. Davis, P. D. Leeson and T. Oprea, *Angew. Chem., Int. Ed.*, 1999, **38**, 3743.
5. D. J. Carini, J. V. Duncia, A. L. Johnson, A. T. Chiu, W. A. Price, P. C. Wong and P. B. Timmermans, *J. Med. Chem.*, 1990, **33**, 1330.
6. I. M. McLay, F. Halley, J. E. Souness, J. McKenna, V. Benning, M. Birrell, B. Burton, M. Belvisi, A. Collis, A. Constan, M. Foster, D. Hele, Z. Jayyosi, M. Kelley, C. Maslen, G. Miller, M-C. Ouldelhkim, K. Page, S. Phipps, K. Pollock, B. Porter, A. J. Ratcliffe, E. J. Redford, S. Webber, B. Slater, V. Thybaud and N. Wilsher, *Bioorg. Med. Chem.*, 2001, **9**, 537.
7. A. J. Collis, M. L. Foster, F. Halley, C. Maslen, I. M. McLay, K. M. Page, E. J. Redford, J. E. Souness and N. E. Wilsher, *Bioorg. Med. Chem. Lett.*, 2001, **11**, 693.
8. P. L. Bamborough, A. J. Collis, F. Halley, R. A. Lewis, D. J. Lythgoe, J. M. McKenna,

I. M. McLay, B. Porter, A. J. Ratcliffe and P. A. Wallace, WO 9856788.

9. H. van de Waterbeemd, D. A. Smith, K. Beaumont and D. K. Walker, *J. Med. Chem.*, 2001, **44**, 1313.

10. V. Austel, *Eur. J. Med. Chem.*, 1982, **17**, 9.

11. M. Baroni, S. Clementi, G. Cruciani, N. Kettaneh-Wold and S. Wold, *Quant. Struct.-Act. Relat.*, 1993, **12**, 225.

12. R. A. Lewis, J. S. Mason and I. M. McLay, *J. Chem. Inf. Comput. Sci.*, 1997, **37**, 599.

13. P. M. Dean and R. A. Lewis, *Molecular Diversity in Drug Design*, Kluwer, Dordrecht, Netherlands, 1999.

14. Y. C. Martin, M. G. Bures, E. A. Danaher, J. DeLazzer, I. Lico and P. A. Pavlik, *J. Comput.-Aided Mol. Des.*, 1993, **7**, 83.

15. G. Jones, P. Willett and R. C. Glen, *J. Comput.-Aided Mol. Des.*, 1995, **9**, 532.

16. D. Barnum, J. Greene, A. Smellie and P. Sprague, *J. Chem. Inf. Comput. Sci.*, 1996, **36**, 563.

17. S. D. Pickett, J. S. Mason and I. M. McLay, *J. Chem. Inf. Comput. Sci.*, 1996, **36**, 1214.

18. J. S. Mason in *Moleclar Diversity in Drug Design*, ed. P. M. Dean and R. A. Lewis, Kluwer, Dordrecht, Netherlands, 1999.

19. A. R. Leach, D. V. S. Green, M. M. Hann, D. B. Judd and A. C. Good, *J. Chem. Inf. Comput. Sci.*, 2000, **40**, 1262.

20. M. Hahn, *J. Med. Chem.*, 1995, **38**, 2080.

21. M. Hahn and D. Rogers, *J. Med. Chem.*, 1995, **38**, 2091.

22. R. D. Cramer III, D. E. Patterson and J. D. Bunce. *J. Am. Chem. Soc.*, 1988, **110**, 5959.

23. G. Klebe, U. Abraham and T. Mietzner, *J. Med. Chem.*, 1994, **37**, 4130.

24. J. Greene, S. Kahn, H. Savoj, P. Sprague and S. Teig, *J. Chem. Inf. Comput. Sci.* 1994, **34**, 1297.

25. M. Hahn, *J. Chem. Inf. Comput. Sci.*, 1997, **37**, 80.

26. C.M. Venkatachalam, P. Kirchhoff and M. Waldman, in *Pharmacophore Perception, Development and Use in Drug Design*, ed. O. F. Guner, IUL Biotechnology Series, California, 1999, Chapter 18, p. 341.

27. E. Gancia, G. Bravi, P. Mascagni and A. Zaliani, *J. Comput.-Aided Mol. Des.*, 2000, **14**, 293.

28. M. Pastor, G. Cruciani, I. M. McLay, S. Pickett and S. Clementi, *J. Med. Chem.*, 2000, **43**, 3233.

29. C. Bissantz, G. Folkers and D. Rognan, *J. Med. Chem.*, 2000, **43**, 4759.

30. G. Jones, P. Willett, R. C. Glen, A.R. Leach and R. Taylor, *J. Mol. Biol.*, 1997, **267**, 727.

31. P. Bambourough, http://www.ukqsar.co.uk/spr2001/pbamorough.pdf

32. J. Sadowski and H. Kubinyi, *J. Med. Chem.*, 1998, **41**, 3325.

33. Y. H. Zhao, J. Le, M. A. Abraham, A. Hersey, P. J. Eddershaw, C. N. Luscombe, D. Boutina, G. Beck, B. Sherborne, I. Cooper and J. A. Platts, *J. Pharm. Sci.*, 2001, **90**, 749.

34. J. A. Gratton, M. H. Abraham, M. W. Bradbury and H. S. Chadha, *J. Pharm. Pharmacol.*, 1997, **49**, 1211.

35. D. E. Clark, *J. Pharm. Sci.*, 1999, **88**, 815.

36. S. Ekins, G. Bravi, S. Binkley, J. S. Gillespie, B. J. Ring, J. H. Wikel and S. A. Wrighton, *J. Pharmacol. Exp. Ther.*, 1999, **290**, 429.

37. G. Klopman and H. Zhu, *J. Chem. Inf. Comput. Sci.*, 2001, **41**, 439.

38. R. D. Saiakhov, L. R. Stefan and G. Klopman, *Perspect. Drug Disc. Des.*, 2000, **19**, 133.

39. M. Cronin, *Curr. Opin. Drug Disc. Dev.*, 2000, **3**, 292–297.

40. G. Cruciani, P. Crivori, P.-A. Carrupt and B. Testa, *Theochem*, 2000, **503**, 17.

41. P. Crivori, G. Cruciani, P.-A. Carrupt and B. Testa, *J. Med. Chem.*, 2000, **43**, 2204.
42. M. J. De Groot, M. J. Ackland, V. A. Horne, A. A. Alex and B. C. Jones, *J. Med. Chem.*, 1999, **42**, 4062.
43. S. D. Pickett, I. M. McLay and D. E. Clark, *J. Chem. Inf. Comput. Sci.*, 2000, **40**, 263.
44. J. M. McKenna, F. Halley, J. E. Souness, I. M. McLay, S. D. Pickett, A. J. Collis, K. Page and I. Ahmed, *J. Med. Chem.*, 2002, **45**, 2173.

The Impact of Physical Organic Chemistry on the Control of Drug-like Properties

Andrew M. Davis and Robert Riley

DEPARTMENT OF PHYSICAL AND METABOLIC SCIENCE,
ASTRAZENECA R&D CHARNWOOD, BAKEWELL ROAD,
LOUGHBOROUGH, LEICESTERSHIRE LE11 5RH, UK

1 Introduction

Physical organic chemistry is the quantitative study of organic reactivity. The work of Hammett in the 1930s and 1940s represented the highlight of this discipline.[1] Hammett recognised that the effect of electron-withdrawing substituents on the rates of reaction and position of equilibria were a property of the substituent and not the reaction. Hammett used the difference in the pK_as of *m,p*-substituted benzoic acids to quantify the electronic effect of substituents, which was termed $\sigma(\sigma = pK_a \text{ H-benzoic acid} - pK_a \text{ X-benzoic acid})$. The sensitivity of a particular reaction to the electronic effect is represented by the reaction constant ρ. This seminal work led to the development of the whole field of physical organic chemistry, dissecting the role of electronics in organic reactivity. So, how does this body of work impact on drug discovery? Taking the lead from physical organic chemistry, many scientists wished to identify similar structure–activity relationships with biological activity. While one can point to a few examples of successful quantitative structure–activity relationships (QSARs) from Hammett's time forwards, the pharmaceutical industry had to wait until the early 1960s to have their own 'biological Hammett' equation with the work of Hansch and Fujita.[2] Their breakthrough was to take a multivariate approach to correlating structure changes with biological activity – using *n*-octanol–water partition coefficients to model the hydrophobic effect, Hammett's σ or Taft's[3] σ^* to represent the electronic effect of substituents and Tafts' *Es* parameter to model steric effects of substituents. Hansch's choice of *n*-octanol–water partition coefficients to model hydrophobicity was inspired, and built upon work of Overton[4] and Meyer[5] stretching back to the turn of the previous century, which had used oil–water partition coefficients to model narcosis. The advantage to Hansch of *n*-octanol was that many drugs will dissolve in it, and its UV transparency made developing experimental methods for determining partition coefficients straight-forward. Hansch and Fujita managed to identify many 100s of such QSARs, which launched the whole area of QSAR upon the pharmaceutical industry.[6] This was the 1960s revolution – and one can imagine that this must have had a similar impact on drug discovery to the growth of computational chemistry in the 1980s and combinatorial chemistry in the 1990s.

Many such correlations were identified, and the fact that ρ, σ and Es could be tabulated meant correlations could be generated and predictions made without any actual measurements.[7] The partitioning system on which the hydrophobicity scale was based became standardised upon n-octanol–water. The compilation of databases containing many 10 000s of measurements allowed algorithms to be developed to predict logPs. LogPs for quite complex drug-like molecules could be predicted with an acceptable degree of accuracy, and these algorithms remain largely unchanged today.[6–8] But the new science of the 1970s–1980s was com-putational chemistry. The rapid increase in affordable computational resources and the development of visualisation algorithms and high-resolution graphics terminals started its own revolution. The focus upon electronic distribution, conformational analysis along with the growth of protein crystallography and molecular docking initiated true rational structure based design.

One of the best examples of structure based design is the design of cyclic urea inhibitors for the enzyme HIV protease.[9] HIV protease is a C_2 symmetric dimer with two aspartate residues in the floor of the active site, one from each of the monomers, and an active site water hydrogen bonded to two flaps in the roof of the active site.

The Dupont-Merck group suggested, based on molecular docking studies, that C_2 symmetric dimers would be potent inhibitors. A combination of molecu-lar docking, pharmacophore definition and database searching first led them to design cycloheptanone-diols, to form two charged reinforced hydrogen bonds with the aspartates, and a ketone to displace the water hydrogen bonded to the flaps (Figure 1). The ketone was replaced with a urea to strengthen the hydrogen bonds to the flaps. Their modelling studies allowed them to optimise hydropho-bic interactions and also correctly predict the stereochemical preferences of the ligands. These compounds were found to be very potent inhibitors of HIV protease, and allowed Dupont-Merck to take DMP323 into development as a possible treatment for AIDS in 1994. This first clinical candidate was rapidly terminated due to variable bioavailability, poor solubility and metabolic insta-bility of the hydroxyl functions. This was replaced by DMP450 with better solubility and good bioavailability in man. This is not the end of the story for this project and we will return to this later.

Two other important developments in pharmaceutical research occurred in the late 1980s/early 1990s, which changed the way we do drug discovery. The first was high throughput screening (HTS) and the second was the appreciation of the influence of drug metabolism and pharmacokinetics (absorption, distribu-tion, metabolism and elimination studies, ADME) within pre-clinical projects. Pfizer's experience was that HTS was not necessarily delivering robust projects into discovery. Chris Lipinski,[10] in his evaluation of Pfizer's 'success' in generat-

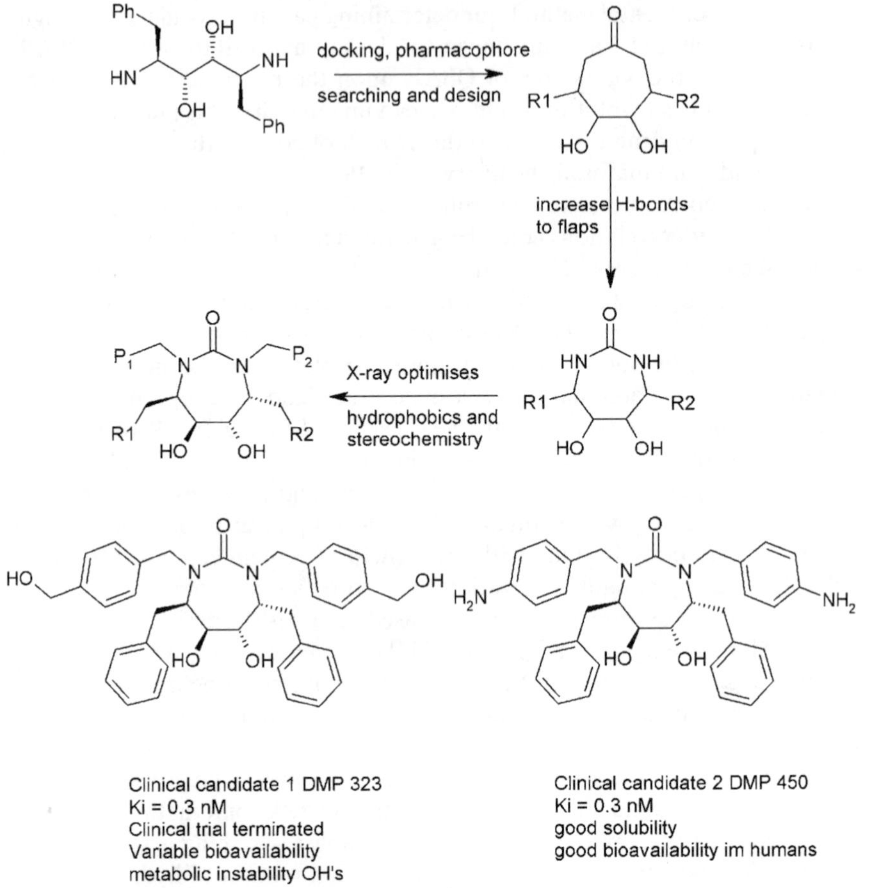

Figure 1 *Dupont-Merck optimisation of a lead for HIV-protease utilising tools of structure-based design – design of cyclic ureas*

ing projects from high throughput screening, found that hits from HTS tended to be lipophilic, and that recent entries into Pfizer's development portfolio were even more lipophilic. In his study of the physical properties of development drugs he proposed that a drug-like property range exists. Of the compounds in development, 90% tend to have $\log P < 5$, MW < 500, number of donors < 5 and number of Ns and Os < 10, which has become known as The Rules of 5. Lipinski's work put physical organic chemistry back on the map – not only in terms of calculated physical properties but also the equal importance of making measurements in high throughput mode – by developing a high throughput solubility screen. Pfizer also introduced registration alerts that have been copied by many pharmaceutical companies – in an effort to improve the properties of compound synthesised and submitted for testing. The importance of this on the way we conduct research has not been lost. Indeed, it has spawned a number of other important papers attempting to classify the drug-like space to allow database screening and library design.[11,12]

It has been reported that 31% drugs in clinical development fail due to pharmacokinetic deficiencies.[13] More recent figures suggest the proportion of failures (1994–1997) may remain as high as 24%.[14]

Our own work used a different database – rather than focus upon development compounds we focused upon marketed oral drugs only. While 90% of the oral drugs lie within the rules of 5 the mean properties of drugs on the market lie far below these property extremes.[15]

The introduction of ADME studies into pre-clinical research has had a huge impact upon the process of drug discovery. It has moved discovery projects away from a focus solely on potency and selectivity to optimising upon two key endpoints: dose to man and dose frequency. The dose-to-man calculations, of which one simple form of the equation is shown in equation 1, provide a useful platform to balance the importance of potency, protein binding, clearance and bioavailability in structure optimisation.

$$Dose = (C_{ss\,av} \times \tau \times Cl)/F \tag{1}$$

where $C_{ss\,av}$ = average plasma concentration required to drive efficacy, τ = dose frequency, Cl = prediction of human clearance and F = predicted human bioavailability.

A more detailed dose-to-man equation is illustrated in equation 2. This equation allows us to balance the effects of volume half-life and dosing frequency.

$$C_{min,\,ss} = \frac{F.Dose.e^{-k\tau}}{V(1 - e^{-k\tau})} \tag{2}$$

where $C_{ss\,min}$ = minimum plasma concentration required at steady state, V = volume of distribution and k = elimination constant.

Drug discovery is now aimed not just at optimising potency and selectivity – but also lowering clearance, maximising absorption and bioavailability, reducing protein binding (increasing free fraction) and of course increasing potency. These properties are balanced against the required dose frequency, which has $t_{1/2}$ implications as stated above. (In a competitive therapeutic environment, once-a-day dosing is usually required and indeed the top 10 best selling oral drugs are all once-a-day dosing.) A good drug could be obtained by having the appropriate combination of all these properties. A selective compound with relatively lower potency may still be acceptable if clearance is lower and/or protein binding is also lowered. However, we still need to consider the total daily dose with respect to adverse drug reactions and perhaps hepatic-enzyme induction.[16,17] We also want to minimise any strong inhibition of the cytochrome P450 enzymes since P450-dependent oxidation is the major mechanism of clearance of many drugs.[18] In modern medicinal therapy regimens drugs are rarely dosed alone but often in cocktails with other drugs – P450 inhibition may lead to significant drug–drug interactions, which may lead to toxicity. Indeed, it has been estimated that up to 3% of hospital admissions may be a result of drug–drug interactions,[18] the most common of which involve CYP metabolism.[19] This seems to make the process of drug optimisation a much more challenging process than in the past – but the rewards would be a more robust nomination candidate, which may be easier and faster to develop, and less likely to fail in the early stages of drug development. Not only are we challenged to produce better quality drug candidates in today's drugs discovery programs, but also better quality drug candidates much faster than ever before. Ideally we would want to build in good ADME properties in the first instance, but the question then remains – how can this be achieved?

2 The Control of ADME Properties

Ideally we would like to have a clear enough understanding of the structural features/properties that control absorption, free fraction, cytochrome P450 inhibition, metabolism, distribution and elimination in order that ideal properties could be designed into molecules before synthesis. It is apparent that many of these important properties are much more dependent upon bulk physical properties than the target SAR.

Absorption

Prediction of structural features that control human intestinal permeability has been a major focus for many years. A number of different experimental protocols have been used to model absorption, including rate of uptake from everted rat gut-loop, Ussing chamber measurements, and more recently permeability across epithelial cell monolayers in culture.[20] The *in vitro* Caco-2 screen has become almost a standard protocol in all pharmaceutical companies and a quantitative model of permeability. The Caco-2 cell line is a human colonic carcinoma cell-line, but it is thought to be a reasonable model of human small intestinal permeability. It contains many uptake and efflux pumps, many of which have been well investigated. However, the relevance of these transporter proteins to human absorption *in vivo* is not that well defined. Even with all these caveats, predictive physicochemical QSAR models have been developed from such screens based on simple physicochemical descriptors.[21–24] QSARs have been developed that highlight the importance of log*P*, hydrogen bond donor counts, molecular weight and polar surface area, amongst others. The problem facing chemists is that many of these studies are based on only small datasets, with very limited chemical diversity, and the relative balance of these properties throughout all of the drug-like space is difficult to determine. Large datasets, that may not exist in the public domain, provide the best chance of obtaining such global models. They will surely contain these descriptors of hydrophobicity, hydrogen bond counts, and some descriptor for molecular size – although of course they may differ in terms of the exact descriptors chosen to represent these underlying physical properties. For instance, although polar surface area is very popular as a descriptor, and even a surrogate for permeability, it is computationally quite an involved calculation. But polar surface area is largely just another way of counting Ns and Os as described by Lipinski's rules of 5.

The mechanism of absorption for the majority of oral drugs is assumed to be passive transcellular permeability.[23] With the Caco-2 screen containing many transporters expressed to differing extents and of unknown *in vivo* relevance, this has prompted some companies to question its usefulness as a screen for passive transcellular permeability, and as a source of data upon which to generate QSAR models. For instance HoffmanLaRoche over the past few years have developed a purely physical-organic model of transcellular permeability as a surrogate for the Caco-2 screen.[24] Instead of having a living cell monolayer separating an aqueous donor and receiver compartment, Roche's PaMPA screen uses a phospholipid membrane soaked pad. Permeation from the apical to basal receiver compartment can therefore only occur by passive transmembrane permeability. If the project aim is to target passive transcellular permeability for oral absorption, then QSAR models based upon the PaMPA screen data may actually produce better predictive models than those based upon the Caco-2 screens. Some companies are so confident in the predictive ability of their computational models for oral permeability that they have actually ceased screening for permeability, and have replaced these experimental protocols with purely computational models.

Lipinski also highlights the importance of molecular weight in drug-likeness. Molecular weight sometimes occurs as a descriptor for oral absorption also. This is one of the more controversial properties, as it is intuitively difficult to understand why molecular weight should or could be a controlling factor. Molecular weight is a correlate of molecular volume, which controls molecular diffusion, which may modulate permeability. Further, clearance by biliary excretion is thought to be dependent upon molecular weight, with the cut-off for rat thought to be 350 and 500 for human.[25] In AstraZeneca R&D Charnwood's own pharmacokinetic database, we can clearly determine a dependence of rat bioavailability upon molecular weight, after we have removed the confounding effects of

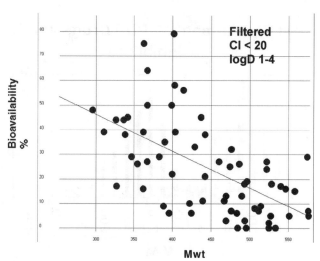

Figure 2 *Dependence of bioavailability in rat upon molecular weight for low clearance compounds (Cl < 20) and intermediate lipophilicities (logD = 1–4)*

clearance and $\log D_{7.4}$ (Figure 2). Whether biliary clearance or permeability is the major reason for the importance of molecular weight, it is clear that molecular weight is an important controlling property. Dupont-Merck[26] found that their cyclic ureas in dog only achieved significant exposure when the molecular weight of their ligand was below 600.

Many pharmaceutical companies had research programmes aimed at inhibition of renin as a possible therapy for hypertension. The X-ray crystal structure of human renin became available in 1989;[27] this also proved the opportunity to exploit the then new computational tools and attempt a rational structure based drug design approach. But to date no renin inhibitor from any of these programs have made it past phase II clinical development. All renin inhibitors in development failed due to problems in achieving sufficient bioavailability due to the high molecular weights of the designed inhibitors and also the cost of production of these complicated molecules.[28] (Figure 3). This is also not the end of this story and we will return to renin later.

Plasma Protein Binding

The degree of plasma protein binding of a compound *in vivo* is as important as its inherent potency in determining *in vivo* efficacy. Efficacy is presumed to be driven by free concentration of drug – not its total concentration. Plasma protein binding is largely controlled by lipophilicity, and reducing lipophilicity causes a concomitant reduction in protein binding. Acids show much higher plasma protein binding than neutrals or bases (Figure 4), but within a homologous series it is still modulated by lipophilicity.

The role of protein binding is nicely illustrated in the discovery of NMDA antagonists by Merck.[29] In optimising their $6.5\,\mu M$ lead I, they managed to

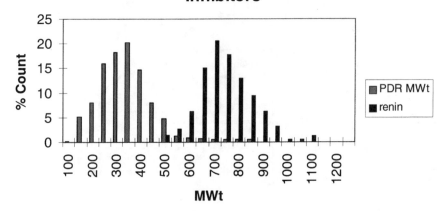

Figure 3 *Comparison of molecular weights of renin inhibitors taken into development with the Physicians Desk Reference Oral Drugs Profile*

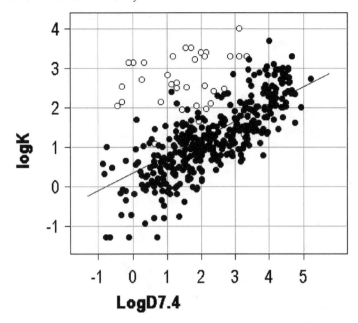

Figure 4 *Plot of log(bound/free) vs. logD₇.₄ for neutrals/bases and acids from AZ R&D Charnwood database*

increase potency almost 3000-fold by introducing a hydrophobic binding motif. *In vivo* the efficacy of II is only 20-fold higher than I. In fact they obtained a good correlation between *in vitro* potency and *in vivo* efficacy (Figure 5). The large loss in efficacy was solely due to the concomitant increase in protein binding with structural optimisation. The difficulty with this lead optimisation project arose because the binding to the receptor in this series of compounds was also largely controlled by hydrophobicity, along with plasma protein binding (Figure 6).

In attempting to derive QSARs for metabolic properties *in vivo*, a number of authors have tried to correct for the confounding role of protein binding by deriving so-called intrinsic or unbound values. For instance, as *in vivo* clearance is modulated by the unbound fraction in plasma (f_u), $\log(Cl/f_u)$ *vs.* logD correlations are often used to demonstrate the role of bulk properties. However, because of the intercorrelation of free fraction with logD, a number of these apparent correlations are completely spurious. This can make deriving an understanding from the literature rather difficult.[30]

3 Volume of Distribution

Volume of distribution not only has a direct role in modulating $t_{1/2}$, but also in determining the difference between C_{max} and C_{min} during the dosing interval. This may be an important consideration in managing the therapeutic margin between efficacy and side effects. From a consideration of the literature, it is clear

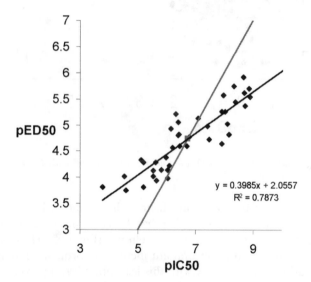

IC$_{50}$ = 6.5 μM
ED$_{50}$ = 52 μmol/kg

IC$_{50}$ = 2.2 nM
ED$_{50}$ = 2.6 μmol/kg

y = 0.3985x + 2.0557
R^2 = 0.7873

pED50

pIC50

Figure 5 *Plot of pED$_{50}$ (in vivo efficacy) vs. pIC$_{50}$ (in vitro binding) for Mercks NMDA antagonists*

that the volume of distribution of a drug is largely governed by its physicochemical properties, most notably charge and lipophilicity.[31] A useful relationship for the volume of distribution at steady state, V_{ss},[32] is given below:

$$V_{ss} = V_p + V_T K_p = V_p + f_u/f_{uT} V_T$$

where V_P = volume plasma compartment, V_T = volume tissue compartment, K_P = plasma : tissue partition coefficient, f_u = fraction drug unbound in plasma and f_{uT} = fraction drug unbound in tissue.

It can be deduced from this that for compounds with large volumes of distribution, $V_{ss} \propto f_u$. Similarly, for compounds with small volumes of distribution, tissue penetration (as reflected in the term, K_p, the tissue to plasma concentration ratio) is limited. Therefore, acidic drugs exhibit volumes of distribution ~ 0.1–0.2 L kg^{-1} as a consequence of extensive plasma protein binding and poor tissue affinity. It can also be deduced that, for such compounds, improvements in V_{ss} will only be made once f_{uT} is addressed, which would entail reducing protein

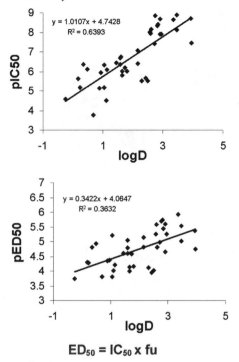

Figure 6 *Plot of pED$_{50}$ (in vivo efficacy)* vs. *logD and pIC$_{50}$ (in vitro binding)* vs. *logD for Merck's NMDA antagonists*

binding while increasing or maintaining tissue affinity. By contrast, basic compounds with moderate lipophilicity will exhibit large volumes of distribution owing to their higher tissue affinity/membrane interactions,[33] and low protein binding relative to their tissue affinity. In general, neutral compounds may be expected to have moderate V_{ss} values since both f_u and f_{uT} may be influenced similarly by lipophilicity. The effects of these considerations can clearly be seen in the AZ R&D Charnwood pharmacokinetic database comparing volumes of distribution of neutrals and basic compounds *vs.* logD (Figure 7). The importance of the balance between tissue affinity and protein binding not only determines the extent of distribution, as measured by the volume of distribution, but also the sites of drug disposition, with acids and neutrals largely distributing to adipose tissue, while bases largely distribute to 'lean' tissue.[34,35]

K_p can be measured from *in vivo* studies and from *in vitro* measurements of binding to tissues and plasma. A reasonable relationship between K_p *in vivo versus in vitro* has been established for eleven drugs by Schumann.[36] However, for several highly cleared compounds there was a discrepancy, suggesting either a flaw in experimental design or active uptake or secretion of such compounds by some tissues. The labour-intensive nature of such measurements and the requirements for human tissue preclude their use as routine screening tools. However, the scope for optimising V_{ss} in order to achieve pharmacokinetic and phar-

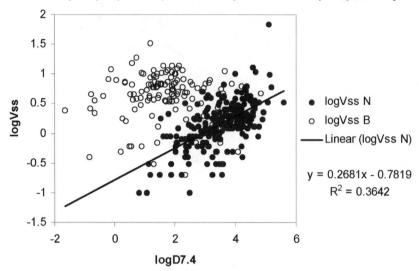

Figure 7 *Plot of logV$_{ss}$ vs. logD$_{7.4}$ for neutral and basic compounds taken from AZ R&D Charnwood pk database. The volume of distribution of neutral compounds clearly follows logD while the volumes of basic compounds do not*

macodynamic duration may be limited by the SAR for the pharmacodynamic target. Critical ionic interactions with amino acid residues may hamper efforts to introduce a basic centre within the molecule, to increase K_p, and may explain why such case histories (*e.g.* amlodipine and rifabutin) are limited and often appear to reflect serendipity arising from a desire to introduce more soluble functionalities.[37]

4 Metabolic Stability and Cytochrome P450 Inhibition

Optimisation of metabolic stability is now accepted to be an equally important objective of a lead optimisation project as optimisation of potency and selectivity. In simple terms, clearance and absorption determine bioavailability and both clearance and the volume of distribution may, at least in theory, be optimised to provide the required elimination half-life. In turn, clearance can be considered to be composed predominantly of renal or hepatic processes.

Passive renal filtration usually occurs with water-soluble compounds with logD$_{7.4}$ < 0 and can be predicted well using allometric scaling, a knowledge of the glomerular filtration rate and the fraction unbound across species.[39] Needless to say, such polar compounds are rarely encountered early in the drug discovery process for targets aimed at chronic oral delivery. Therefore, hepatic clearance tends to dominate either *via* biliary clearance of unchanged drug or enzyme-catalysed biotransformation (metabolism).

The lack of detailed knowledge of species similarities or differences in uptake or efflux transporter proteins makes optimisation of hepatic uptake/biliary clearance a less attractive proposition at present, despite initial allusions to molecular weight thresholds and inter-species trends in biliary elimination.[40]

In order to facilitate the lead optimisation process, it may therefore seem prudent to select a series of compounds for which the rate-determining processes for the ADME fate are well understood. Our lack of knowledge of species differences, structure–activity relationships, tissue distribution and relative activity of individual phase 2 enzymes suggests that it may be prudent to target metabolic clearance *via* CYP oxidation, where appropriate. Many publications are appearing defining pharmacophores for cytochrome P-450 dependent oxidation based on homology modelling-docking, QSARs based upon recursive partitioning and molecular orbital calculations.[41–47] These may point towards specific structural features that lead to molecular recognition and subsequent metabolism, and hence guide a discovery program towards structural modification to block these recognition features or metabolic sites.

Optimisation of bulk properties may also be a successful optimisation strategy in its own right. We have recently analysed data from AstraZeneca R&D Charnwood's own rat hepatocyte database (Figure 8). This data comes from many projects and structural classes. It clearly demonstrates that in a low log*D* range 70% of compounds tend to be stable, with only a small percentage with high intrinsic clearances. In contrast as log*D*s of compounds rise to 1–3 and further to 3–5 we see a decrease in the percentage of stable compounds and an increase in the percentage of highly unstable compounds. In this large dataset of structurally diverse compounds we can easily demonstrate an underlying control of hydrophobicity. One possible reason for this observation is that for many compounds the major route of clearance is through oxidation by CYP-3A4.[48] This is probably the most non-specific of all the cytochrome P450 enzymes and shows little structural specificity; recognition is largely controlled by hydrophobicity. This is true for substrates and inhibitors of 3A4, as has been recently demonstrated by the simple QSAR for structurally diverse 3A4 inhibitors, which highlights the importance of hydrophobicity, and the increase in affinity afforded

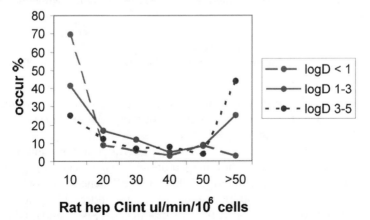

Figure 8 *Analysis of intrinsic clearances in rat hepatocytes from AZ R&D Charnwood database grouped by logD and by intrinsic clearance in rate hepatocytes*

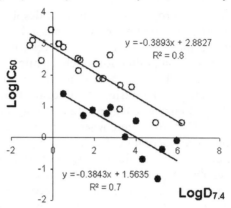

Figure 9 *Plot of pK$_i$ vs. logD for a diverse set of CYP-3A4 inhibitors: dark points – neutral nitrogen acceptor compounds, e.g. pyridine, imidazole, light points – compounds that do not contain a neutral nitrogen acceptor*

by the presence of a neutral nitrogen acceptor in the molecules[49] (Figure 9).

Lipinski's rules of 5 focused upon properties for oral absorption. But these properties are of more widespread importance than this. As has been highlighted in this review, hydrophobicity, polar atom counts, molecular weight and charge type are also of fundamental importance in controlling protein binding, cytochrome inhibition and metabolic stability, volume of distribution, sites of disposition and blood–brain barrier permeability,[50–53] renal excretion and, although not well understood, biliary excretion. The conundrum for many projects is that while many of the ADME properties improve by modulating these bulk properties, potency is also dependent upon the same properties, and particularly hydrophobicity (Figure 10). There are then three possible optimisation targets:

1. To find a new receptor interaction that increases potency, while modulating these bulk properties
2. Find a position for chemical modulation that can be used to improve the bulk properties while maintaining receptor potency
3. Trade off potency for an improvement in DMPK properties by focusing on dose-to-man calculations

While these sound simple project aims, in reality they may prove very difficult to achieve. It is also true that drugs can be found at the extremes of the bulk property space. For instance the HIV protease inhibitors saquinavir and nelfinavir, or the immunosuppressant cyclosporin, do not represent ideal role models for oral programs. They may represent first in class drugs and are clinically financially successful but they have DMPK deficiencies, of variable or poor bioavailability,[54,55] that potentially could be addressed by a follow-up compound that would then easily become best in class. Speed and quality in the pharmaceutical industry now require your candidate to be a potential first and best in class drug. For example although nifedipine was first in class dihyd-

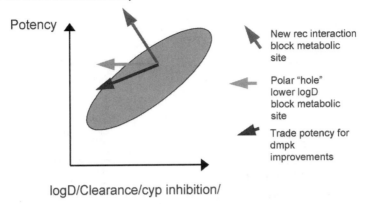

Figure 10 *The drug design conundrum – maintaining potency while optimising DMPK properties*

ropyridine calcium channel antagonist, the once-a-day amlodipine from Pfizer is the best in class drug. We are challenged with always discovering the new 'amlodipine' first!

Lipinski's seminal work in defining a drug-like space probably had its biggest influence in lead generation. The growth of HTS in the 1990s stimulated the development of high throughput chemical synthesis, as a means of feeding HTS screens with more chemical targets. Many early combinatorial chemistry libraries have since been discarded and the design of drug-like libraries became the new paradigm. But recently a number of groups have questioned the drug-like paradigm for lead discovery.[56,57]

Experience tells us that often chemical optimisation leads to an increase in molecular weight and lipophilicity of lead structures. In this case optimisation of drug-like starting points would take the bulk properties outside the drug-like window. The logical conclusion of this thought experiment would be that the libraries for lead generation should be targeted at properties somewhat more conservative than drug-like properties. The starting point should not be drug-like necessarily but lead-like, and tend to have lower molecular weight and lipophilicity – allowing room for chemical optimisation.

Dupont-Merck in 1994 took DMP450 into clinical development as the second clinical compound with improved bioavailability and solubility. But this was not the end of the story – after reaching patients they found this compound had only 'modest' potency.[58] The company took the project back to the research phase with a largely similar discovery program – with the addition of a potency screen in the presence of plasma protein, as too high plasma protein binding was identified as the main shortcoming of DMP450. The latest development candidate (Figure 11), with maintained potency and substantially reduced protein binding, is DMP850 and it will be fascinating to follow the development of this ongoing drug discovery story.

If any protein target was likely to succeed in being druggable it should have been renin. With most of the world's pharmaceutical companies at some point

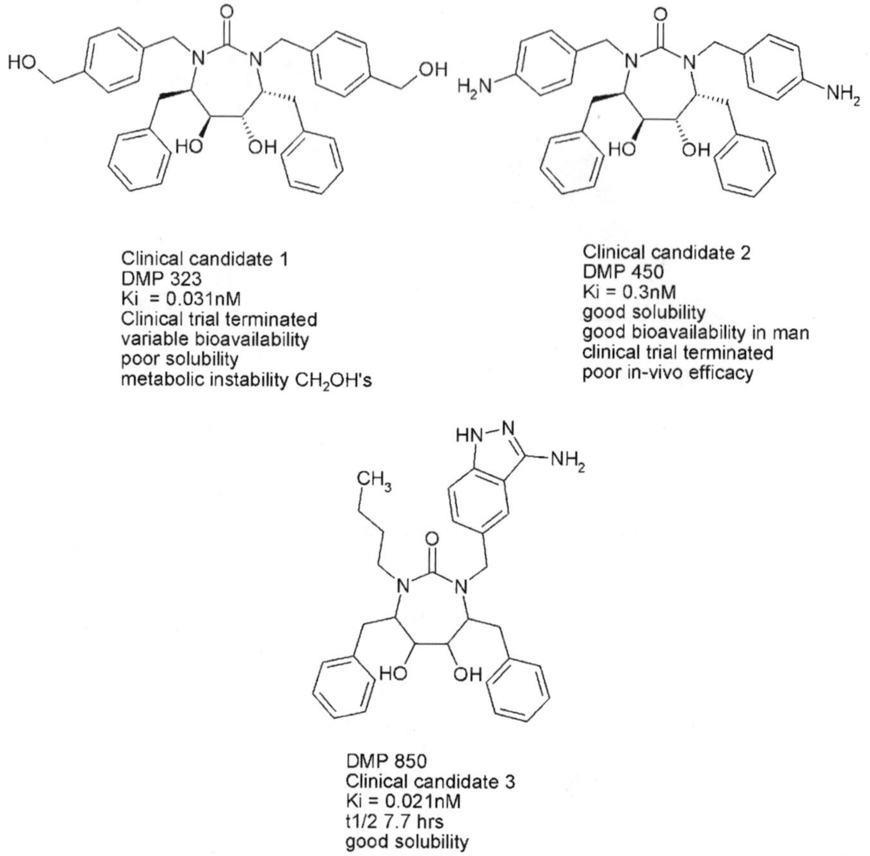

Clinical candidate 1
DMP 323
Ki = 0.031nM
Clinical trial terminated
variable bioavailability
poor solubility
metabolic instability CH₂OH's

Clinical candidate 2
DMP 450
Ki = 0.3nM
good solubility
good bioavailability in man
clinical trial terminated
poor in-vivo efficacy

DMP 850
Clinical candidate 3
Ki = 0.021nM
t1/2 7.7 hrs
good solubility

Figure 11 *Development of Dupont-Merck's cyclic urea HIV protease inhibitors by optimising (lowering) protein binding*

having a renin inhibitor discovery program and with the availability of high-resolution X-ray structures to guide compound design, by now it should have yielded a drug. As we highlighted earlier renin has so far not yielded its prize. But companies have not given up on this target yet. Recently Roche have described compound Roche 2, a nM renin inhibitor developed from a 22 μM HTS hit[59] (Figure 12). At MW 550 it is one of the smallest most drug-like ligands to have been discovered, and is furthest from the peptidic ligands of the 1990s. Interestingly both the lead and potent ligand induce a major conformational change in renin active site, never previously observed in all the years of X-ray crystallographic studies, and which in itself represents a unique new starting point for drug-design programs.[15] Again, this is clearly not the end of the renin story, and it will be fascinating to see how modern discovery programs deal with this old target.

Physical organic chemistry is now back in vogue. As well as its traditional role in Hansch analysis in the understanding of drug–receptor interactions, the role

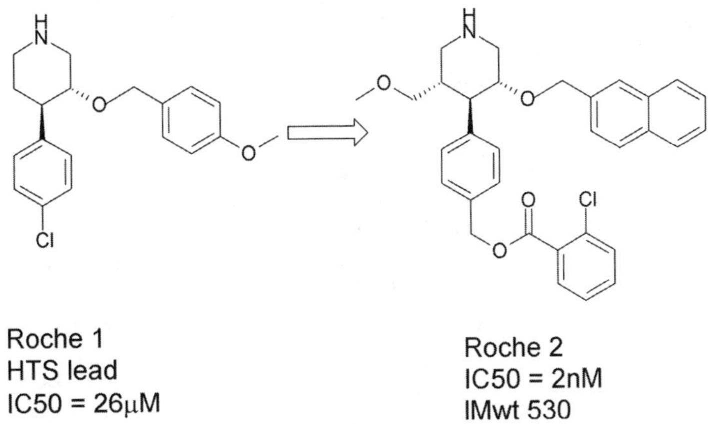

Roche 1
HTS lead
IC50 = 26μM

Roche 2
IC50 = 2nM
IMwt 530

Figure 12 *Roche 'lower' molecular weight renin Inhibitor – development from a lead-like HTS screening hit*

of hydrophobicity, size, and electronic influences are also important in the area of ADME optimisation. A focus upon Lipinski type properties and the efficient control of these aids more than just absorption, and can help move projects from ligand optimisation towards true drug design.

References

1. L.P. Hammett, *Physical Organic Chemistry*, McGraw-Hill, New York, 1940 and 1970.
2. C. Hansch, P.P. Maloney and T. Fujita, *Nature*, 1962, **178**, 4828.
3. R.W. Taft, *Separation of Polar, Steric, and Resonance Effects in Organic Chemistry*, ed. M.S. Newman, Wiley, New York, 1956.
4. E. Overton, *Z. Phys. Chem.*, 1897, **22**, 189–209.
5. H. Meyer, *Arch. Exp. Pathol. Pharmakol.*, 1899, **42**, 109–118.
6. C. Hansch and W.J. Dunn, III, *J. Pharm. Sci.*, 1972, **61**, 1; C. Hansch and J.M. Clayton, *J. Pharm. Sci.*, 1973, **62**, 1–21.
7. C. Hansch and A. Leo, *Substituent Constants for Correlation Analysis in Chemistry and Biology*, Wiley Interscience New York, 1979; J. Chou and P. Jurs, *J. Chem. Inf. Comput. Sci.*, 1979, **19**, 172.
7a. R. Rekker, *The Hydrophobic Fragmental Constant*, Elsevier, Amsterdam, 1977.
8. Advanced Chemistry Development Inc. 90 Adelaide Street West, Suite 702, Toronto, Ontario M5H 3V9, Canada.
9. P.Y.S. Lam, P.K. Jadhav, C.J. Eyermann, C.N. Hodge, Yu Ru, L.T. Bacheler, O. Meek, M.J. and M.M. Rayner, *Science*, 1994, **263**, 380–384.
10. C.A. Lipinski, F. Lombardo, B.W. Dominy and P.J. Feeney, *Adv. Drug Del. Rev.*, 1997, **23**, 2–25.
11. Ajay, W.P. Walters and M.A. Murcko, *J. Med. Chem.*, 1998, **41**, 3314–3324; Ajay, W.P. Walters and M.A. Murcko, *Curr. Opin. Chem. Biol.*, 1999, **3**, 384–387.
12. J. Sadowski and H. Kubinyi, *J. Med. Chem.*, 1998, **41**, 3325.
13. R.A. Prentis, Y. Lis and S.R. Walker, *Br. J. Clin. Pharmacol.*, 1988, **25**, 387–396.

14. N. McAuslane, *Accelerating preclinical development*, Vision in Business, Conference, Nice, France, 25–26 February 1999.
15. S.J. Teague and A.M. Davis, *Angew. Chem. Int. Ed.*, 1999, **38**, 3743.
16. J. Uetrecht, *Curr. Opin. Drug Disc. Dev.*, 2001, **4**, 55.
17. D.A. Smith, *Eur. J. Pharm. Sci.*, 2000, **11**, 185.
18. C.A. Jankel and L.K. Fitterman, *Drug Safety*, 1993, **9**, 51.
19. H. Chiba Chiryou, *J. Ther.*, 1994, **76**, 2214.
20. P. Artursson and J. Karlsson, *Biochem. Biophys. Res. Commun.*, 1991, **175**, 880–885.
21. A. Tsuji and I. Tamai, *Pharm. Res.*, 1996, **13**, 963–977.
22. Y.-L. He, S. Murby, L. Gifford, A. Collett, G. Warhurst, K.T. Douglas, M. Rowland and J. Ayrton, *Pharm. Res.*, 1996, **13**, 1673–1678.
23. R.A. Conradi, A.R. Hilgers, N.F. Ho and P.S. Burton, *Pharm. Res.*, 1991, **8**, 1453–1460.
24. K. Palm, K. Luthman, A.-L. Ungell, G. Strandlund, F. Beigi, P. Lundahl and P. Artursson, *J. Med. Chem.*, 1998, **41**, 5382–5392.
23. O.H. Chan and B.H. Stewart, *Drug Disc. Today*, 1996, **1**, 461.
24. M. Kansy, F. Senner and K. Gubernator, *J. Med. Chem.*, 1998, **41**, 1007–1010.
25. R.M.J. Ings, in *Medicinal Chemistry: Principles and Practice*, ed. F.D. King, Royal Society of Chemistry, Cambridge, 1994, p. 78.
26. G.V. De Lucca and P.Y.S. Lam, *Drugs Future*, 1998, **23**, 987–994.
27. A.R. Sielecki, K. Hayakawa, M. Fujinaga, M.E. Murphy, M. Fraser, A.K. Muir, C.T. Carilli, J.A. Lewicki, J.D. Baxter and M.N. James, *Science*, 1989, **243**, 1346.
28. M.A. Navia and P.R. Chaturvedi, *Drug Disc. Today*, 1996, **1**, 179–189.
29. M. Rowley, J.J. Kulagowski, A.P. Watt, D. Rathbone, G.I. Stevenson, R.W. Carling, R. Baker, G.R. Marshall, J.A. Kemp, A.C. Foster, S. Grimwood, R. Hargreaves, C. Hurley, K.L. Saywell, M.D. Trickelbank and P.D. Leeson, *J. Med. Chem.*, 1997, **40**, 4053–4068.
30. A.M. Davis, D. Salt and P. Webborn, *Drug Met. Disp.*, 2000, **28**, 103.
31. D.A. Smith, B.C. Jones and D.K. Walker, *Med. Res. Rev.*, 1996, **16**, 243–266.
32. J.H. Lin, *Drug Met. Disp.*, 1995, **23**, 1008.
33. R.P. Austin, A.M. Davis and C.N. Manners, *J. Pharm. Sci.*, 1995, **84**, 1180.
34. M. Bickel, *Adv. Drug Res.*, 1994, **25**, 26.
35. P. Barton, A.M. Davis, D.J. McCarthy and P.J.H. Webborn, *J. Pharm. Sci.*, 1997, **86**, 1034–1039.
36. G. Schumann, B. Fichtl and H. Kurz, *Biopharm. Drug Dispos.*, 1987, **8**, 73.
37. H. Waterbeemd, D.A. Smith, K. Beaumont and D.K. Walker, *J. Med. Chem.*, 2001, **44**, 1.
38. S. Kacew, M.J. Reasor and Z. Ruben, *Drug Metab. Rev.*, 1997, **29**, 355.
39. I. Mahmood and J.D. Balian, *Clin. Pharmacokinet.*, 1999, **36**, 1–11.
40. J. Lin, *Drug Metab. Dispos.*, 1995, **23**, 1008.
41. D.F.V. Lewis, M. Dickins, P.J. Eddershaw, M.H. Tarbit and P.S. Goldfarb, *Drug Metab. Drug Interact.*, 1999, **15**, 1.
42. J.P. Jones, M. He, W.F. Trager and A.E. Rettie, *Drug Metab. Dispos.*, 1996, **24**, 1–6.
43. S. Ekins and R.S. Obach, *J. Pharmacol. Exp. Ther.*, 2000, **295**, 463–473.
44. S. Ekins, G. Bravi, S. Binkley, J.S. Gillespie, B.J. Ring, J.H. Wikel and S.A. Wrighton, *Drug Metab. Dispos.*, 2000, **28**, 994.
45. S. Ekins, G. Bravi, S., Binkley, J.S. Gillespie, B.J. Ring, J.H. Wikel and S.A. Wrighton, *Pharmacogenetics*, 1999, **9**, 477.
46. D.A. Smith, M.J. Ackland and B.C. Jones, *Drug Disc. Today*, 1997, **2**, 479–486.
47. D.A. Smith, M.J. Ackland and B.C. Jones, *Drug Disc. Today*, 1997, **2**, 406.

48. R.J. Bertz and G.R. Granneman, *Clin. Pharmacokinet.*, 1997, **32**, 210.
49. R. Riley, A.J. Parker, S. Trigg and C.N. Manners, *Pharm. Res.*, 2001, **18**, 652–655.
50. R.C. Young, R.C. Mitchell, T.H. Brown, C.R. Ganellin, R. Griffiths, M. Jones, K.K. Rana, D. Saunders, I.R. Smith, N.E. Sore and T.J. Wilks, *J. Med. Chem.*, 1988, **31**, 656–671.
51. M.H. Abraham, H.S. Chadha and R.C.J. Mitchell, *Pharm. Sci.*, 1994, **83**, 1257–1268.
52. F. Lombardo, J.F. Blake and W.J. Curatolo, *J. Med. Chem.*, 1996, **39**, 4750–4755.
53. J. Kelder, P.D.J. Grootenhuis, D.M. Bayada, L.P.C. Delbressine and J.P. Ploemen, *Pharm. Res.*, 1999, **16**, 1514.
54. B.D. Kahan, M. Welsh, L. Schoenberg, L.P. Rutzky, S.M. Katz, D.L. Urbauer and C.T. Van Buren, *Transplantation*, 1996, **62**, 599.
55. C.M. Perry and S. Noble, *Drugs*, 1998, **55**, 461–486.
56. S.J. Teague, A.M. Davis, P.D. Leeson and T. Oprea, *Angew. Chem. Int. Ed.*, 1999, **38**, 2743–2748.
57. M.M. Hann, A.R. Leach and G. Harper, *J. Chem. Inf. Comput. Sci.*, 2001, **44**, 856–864.
58. J.D. Rodgers, P.Y.S. Lam, B.L. Johnson, H. Wang, S.S., Ko, S.P. Seitz, G.L. Trainor, P.S. Anderson, R.M. Klabe, L.T. Bacheler, B. Cordova, S. Garber, C. Reid, M.R. Wright, C.-H. Chang and S. Erickson-Viitanen, *Chem. Biol.*, 1998, **5**, 597–608.
59. C. Oefner, A. Binggeli, V. Breu, D. Bur, J.P. Clozel, A. D'Arcy, A. Dorn, W. Fischli, F. Grüninger, R. Güller, G. Hirth, H. Mårki, S. Mathews, M. Müller, R.G. Ridley, H. Stadler, E. Vieira, M. Wilhelm,and F. Winkler, *Wostl. Chem. Biol.*, 1999, **6**(3), 127–131.

Mutagenesis and Modelling Highlight the Critical Nature of the TM2-loop-TM3 Region of Biogenic Amine GPCRs

P. Hunt, J. Stanton, E. Handford, A. Heald, H. Skynner,
M. Knowles, G. McAllister, M. Beer, A. Macleod, L. Street
and J. L. Castro

MERCK, SHARP & DOHME NEUROSCIENCE RESEARCH CENTRE,
TERLINGS PARK, EASTWICK ROAD, HARLOW, ESSEX CM20 2QR,
UK

1 Introduction

The understanding of biogenic amine G-protein coupled receptor structure and function has been developing for over five decades and has grown more and more complex since the first proposal of an adrenergic receptor.[1] The use of more and more refined homology models based on bacterial and bovine versions of rhodopsin have helped to interpret and stimulate experiments into the analysis of these receptors. Molecular cloning techniques have expanded the diversity of known GPCRs and hence have increased the complexity of the problem of selective drug design. However, they have also allowed the examination of the relationships between sequence variation and pharmacology, not only for different receptor subtypes but also for the same receptor in different species. The cloning technology enabled the identification of two 5-HT_{1D} receptor subtypes, $5\text{-HT}_{1D\alpha}$ and $5\text{-HT}_{1D\beta}$ receptors, in several mammalian species.[2] These cloned receptors share similar pharmacological profiles but our interest in these receptors stems from agonism at the $5\text{-HT}_{1D\alpha}$ receptor, which is thought to contribute to the anti-migraine action of the triptan class of compounds.[3] The human $5\text{-HT}_{1D\beta}$ receptor has a similar amino acid identity to the rat 5-HT_{1B} receptor despite their differing pharmacological profiles which subsequently led to the suggestion that the $5\text{-HT}_{1D\alpha}$ and $5\text{-HT}_{1D\beta}$ receptors be re-named 5-HT_{1D} and 5-HT_{1B} receptors respectively.[4]

In general, the affinity of compounds at the 5-HT_{1D} receptor appears to be consistent across species with the possible exception of ketanserin which displaces [³H]5-HT binding to the human, rabbit, guinea pig and rat 5-HT$_{1D}$ receptors with a higher affinity than the dog 5-HT$_{1D}$ receptor. [5,6] However, species differences in the pharmacology of G-protein linked receptor sub-types are known and the residues responsible for these pharmacologies are becoming increasingly apparent, *e.g.* the human and rat 5-HT$_{1B}$ receptors with the substitution of a transmembrane VII asparagine in the rat to a threonine in the human. [7]

In our search for a selective human 5-HT$_{1D}$ receptor agonist and hence an anti-migraine treatment without the potential coronary side effects of the triptan class, we needed to study the affinities of a series of selective, and non-selective 5-HT$_{1D}$/5-HT$_{1B}$ receptor compounds at cloned human, dog and rat 5-HT$_{1D}$ and 5-HT$_{1B}$ receptors, stably expressed in cell lines and at mutant receptors, transiently expressed, to determine where changes in receptor sequence might account for any differences in moving between species.

2 Methods

CHO cells stably expressing the cloned human or dog 5-HT$_{1D}$ and 5-HT$_{1B}$ receptors and rat 5-HT$_{1D}$ receptor, and HeLa cells stably expressing the cloned rat 5-HT$_{1B}$ receptor, were homogenised in ice-cold 50 mM Tris HCl buffer (pH 7.7 at RT) and centrifuged at 48 000 × g, at 4 °C, for 11 min. The resulting supernatant was discarded and the pellet re-suspended in the same volume of ice-cold Tris HCl buffer before being incubated at 37 °C for 10 min, to remove any endogenous 5-HT, and re-centrifuged at 48 000 × g, at 4 °C, for a further 11 min. The final pellet was then re-suspended in 50 mM Tris assay buffer containing 0.1% ascorbate, 10 μM pargyline and 4 mM CaCl$_2$, pH 7.7 at RT, to give 4–6 mg wet weight per tube. All assays were carried out in duplicate. Test drug or buffer was incubated with 500 μl membrane, [³H]5-HT (2.0 nM) in a final assay volume of 1 ml, at 37 °C in a shaking water bath. 5-HT (10 μM) was used to define non-specific binding. The incubation was started by addition of the membrane suspension and was terminated after 30 min by rapid filtration over GF/B filters (pre-soaked in 0.3% polyethylenimine/0.5% Triton X–100) using a Brandel cell harvester. Each assay tube was washed twice with 4 ml of ice-cold Tris HCl buffer. Radioactivity was counted by liquid scintillation spectrometry (45–55% efficiency).

Transient transfection of HEK 293 was achieved by mixing plated cells with a mix of 0.25M CaCl$_2$, 15–20 μg DNA (prepared using the QuikChange Site-Directed Mutagenesis Kit[8] with standard conditions and purified through a Qiagen Maxi-prep system[9]) and BBS buffer in a medium comprising Eagles' modified Minimal Essential Medium (EMEM), heat inactivated foetal calf serum, L-glutamine, penicillin/streptomycin and non-essential amino-acids (NEAA) in a 50:5:1:1:1 volume ratio. The chimeras were generated by cleavage of each sequence at the LYS237-ARG238 bond by the Eco47III restriction enzyme and recombined using standard procedures. [10]

3 Data analysis

Dose–response curves were plotted as percent inhibition *versus* drug concentration and were analysed by non-linear least squares regression analysis using an iterative curve fitting routine (Marquardt-Levenberg method) provided by the data manipulation software RS/1 (BBN Software Products Corporation, Cambridge, MA, USA).

4 Results

The affinity values for compounds at the cloned human, dog and rat 5-HT_{1D} and 5-HT_{1B} receptors are shown in Table 1.

5 Discussion

The sequences of all the species and subtypes are shown in Figure 1 and whilst the amino acid sequence of the cloned human and rat 5-HT_{1B} receptors remains highly conserved, differences in the pharmacology of these receptors are well documented.[7,11–13] Conversely, the highly conserved 5-HT_{1D} receptor (across species) appears, to date, to retain similar pharmacologies. One exception is that of ketanserin which displays a higher affinity for the human, rabbit, guinea pig and rat 5-HT_{1D} receptors over the dog 5-HT_{1D} receptor.[5,6]

In the first part of the study, a series of human 5-HT_{1D} receptor selective and $5\text{-HT}_{1D}/5\text{-HT}_{1B}$ receptor non-selective compounds have been assessed in $[^3\text{H}]5\text{-}$ HT radioligand binding displacement assays to determine their affinities at cloned human, dog and rat 5-HT_{1D} and 5-HT_{1B} receptors. The results are given in Table 1.

The non-selective human $5\text{-HT}_{1D}/5\text{-HT}_{1B}$ receptor ligands sumatriptan, CP 122,288, (NS1) and (NS2) (Figure 2) displayed high affinities for the 5-HT_{1D} receptors in all three species. High affinities were also seen at the human and dog 5-HT_{1B} receptors, but lower affinities were seen, as anticipated, at the rat 5-HT_{1B} receptor.[14] The human 5-HT_{1D} receptor selective compounds (S1), (S2) and (S3), however, whilst displaying high affinities at the human 5-HT_{1D} receptor, yielded relatively low affinities at the dog and rat 5-HT_{1D} receptors. The affinities of these compounds were similar at the human and dog 5-HT_{1B} receptors but again, not unexpectedly, were weak in displacing $[^3\text{H}]5\text{-HT}$ from the rat 5-HT_{1B} receptor. Hence, compounds which displayed $5\text{-HT}_{1D}/5\text{-HT}_{1B}$ selectivity at human receptors were non-selective in the dog. These compounds did retain selectivity in the rat but this was due, in the most part, to their weak binding at the rat 5-HT_{1B} receptor rather than their high affinity at rat 5-HT_{1D} receptor.

Having established these species differences with our selective compounds we attempted to discover the residues within the receptors responsible for this effect. To that end, a number of mutants of the human $5\text{-HT}_{1D}/5\text{-HT}_{1B}$ receptors and dog 5-HT_{1D} receptor were created along with several chimeric dog/human 5-HT_{1D} receptors. Molecular models of the receptors were generated to assist in

Table 1 *Comparison of affinity values for compounds at cloned human, dog and rat 5-HT$_{1D}$ and 5-HT$_{1B}$ receptors*

Compound	Human			Dog			Rat		
	5-HT$_{1D}$	5-HT$_{1B}$	5-HT$_{1D}$ receptor selectivity	5-HT$_{1D}$	5-HT$_{1B}$	5-HT$_{1D}$ receptor selectivity	5-HT$_{1D}$	5-HT$_{1B}$	5-HT$_{1D}$ receptor selectivity
Sumatriptan	8.2±0.1	8.0±0.1	2×	8.2±0.1	7.9±0.1	2×	7.7±0.2	7.2±0.1	3×
CP 122,288	8.3±0.1	8.2±0.2	1×	8.2±0.1	8.0±0.2	2×	8.0±0.1	6.7±0.1	20×
(NS1)	9.5±0.1	8.9±0.1	4×	9.1±0.1	8.8±0.1	2×	9.0±0.2	7.8±0.1	16×
(NS2)	9.0±0.1	8.5±0.2	3×	8.6±0.1	8.7±0.1	0.8×	8.5±0.1	7.7±0.1	6×
(S1)	8.9±0.1	6.6±0.1	200×	7.3±0.2	6.8±0.1	3×	7.4±0.1	5.3±0.1	130×
(S2)	9.8±0.2	7.6±0.2	160×	8.2±0.1	7.9±0.1	2×	8.1±0.2	6.6±0.1	32×
(S3)	9.0±0.1	6.9±0.1	130×	7.5±0.1	7.1±0.1	3×	7.2±0.1	5.1±0.1	130×

Results are expressed as pIC$_{50}$ values ($-\log_{10}$ concentration of drug required to inhibit specific binding by 50%) $\pm$ SEM, $n \geqslant 3$.

The 5-HT$_{1D}$ receptor selectivity is $\dfrac{\text{IC}_{50} \text{ at the 5-HT}_{1B} \text{ receptor}}{\text{IC}_{50} \text{ at the 5-HT}_{1D} \text{ receptor}}$

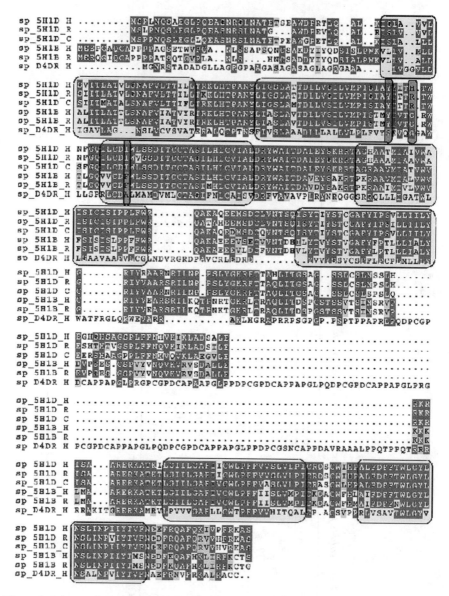

Figure 1 *Pile-up of all six sequences where the species is designated by H – human, R – rat and C – dog. The putative transmembrane domains are coloured and the three residues important in the binding region TYR98, HIS102 and ILE113 are also highlighted*

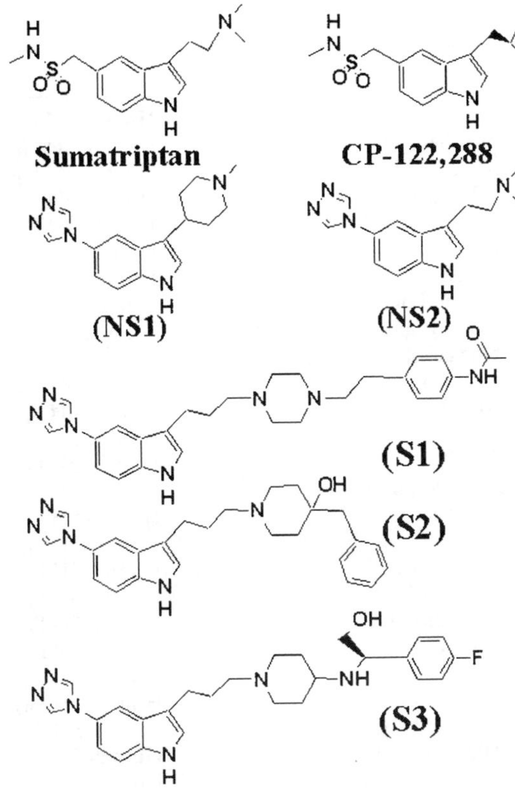

Figure 2 *Selective and non-selective 5HT$_{1D/1B}$ agonists*

the selection of the mutants and in the interpretation of the effects on compound binding affinity which are outlined in Table 2.

6 Molecular Modelling and Discussion

All of the non-selective 5-HT$_{1D}$ /5-HT$_{1B}$ compounds are molecules which only have small substituents on the protonatable nitrogen, with selectivity being gained with larger, aromatic containing, substituents. We assumed that there was no difference in the binding modes of the 5-HT$_{1D}$ /5-HT$_{1B}$ non-selective compounds and 5HT; therefore the indole 5-substituent would bind to the SER residues on Transmembrane Helix 5 (TM 5) in the same manner as the hydroxyl of 5HT and the protonated nitrogen binds to the conserved ASP residue on TM 3 (Figure 3). In this orientation it was most likely that the larger substituents of the selective compounds would extend into the region defined by TMs 1, 2, 3 and 7. The most influential mutant made was the ILE113PHE change which is indeed in this area. This mutant, in the main, did not affect the binding of the non-selective compounds (at most a 2.5 fold shift was seen) yet markedly affected

Table 2 *Mutations listing with receptor/mutant and effect on binding*

Mutation	Outcome	Species
TYR 98 LEU	No radiolabel binding	Human $5HT_{1D}$
ILE 100 VAL	No effect	Human $5HT_{1D}$
HIS 102 GLY	Affected binding	Human $5HT_{1D}$
HIS 102 ARG	Affected binding	Human $5HT_{1D}$
ASN 106 ALA	No radiolabel binding	Human $5HT_{1D}$
GLN 108 GLU	No effect	Human $5HT_{1D}$
ASP 112 ALA	No effect	Human $5HT_{1D}$
ILE 113 PHE	Affected binding	Human $5HT_{1D}$
TRP 114 ALA	No radiolabel binding	Human $5HT_{1D}$
SER 117 ALA	No radiolabel binding	Human $5HT_{1D}$
THR 157 VAL	Affected binding	Human $5HT_{1D}$
ILE 161 LEU	No effect	Human $5HT_{1D}$
SER 166 ALA	No effect	Human $5HT_{1D}$
SER 170 ALA	No effect	Human $5HT_{1D}$
SER 201 ALA	Affected binding	Human $5HT_{1D}$
THR 202 ALA	No effect	Human $5HT_{1D}$
TRP 314 ALA	No radiolabel binding	Human $5HT_{1D}$
PHE 317 ALA	No radiolabel binding	Human $5HT_{1D}$
SER 321 ASP	No radiolabel binding	Human $5HT_{1D}$
ASP 329 ALA	No radiolabel binding	Human $5HT_{1D}$
SER 330 ALA	No effect	Human $5HT_{1D}$
HIS 334 ASP	No effect	Human $5HT_{1D}$
PHE 338 ALA	Affected binding	Human $5HT_{1D}$
THR 342 VAL	Affected binding	Human $5HT_{1D}$
PHE 113 ILE	Affected binding	Human $5HT_{1B}$
ARG 102 HIS	Affected binding	Dog $5HT_{1D}$
Chimera 1	No effect	Dog/Human $5HT_{1D}$
Chimera 2	No effect	Human/Dog $5HT_{1D}$
LEU 112 PHE	Affected binding	Human D4
MET 113 VAL	No radiolabel binding	Human D4

the selective compounds (affinity shifts from 3.5 to 31 fold). Although this change from the $5HT_{1D}$ to $5HT_{1B}$ amino acid reduced the affinities of the selective compounds at the $5HT_{1D}$ receptor, the reverse mutation in the $5HT_{1B}$ receptor did not improve the affinities of the selective compounds. Instead the mutation *reduced* the affinities of both the selective and non-selective compounds (affinity shifts from 4 to 11 fold).

If one examines the region around ILE113 in our homology model of $5HT_{1D}$ (Figure 4) one can see a possible interaction with TYR98, a residue which is conserved across all of the $5HT_1$ family of receptors. In $5HT_{1D}$ there is enough room for the TYR and ILE residues; however, if the ILE is replaced by the larger PHE then the two sidechains would clash. In the $5HT_{1B}$ receptor this clash can be relieved by the movement of the TYR sidechain upwards into the space created by the presence of GLY102 one helical turn above. In $5HT_{1D}$ this GLY102 is a HIS residue and so there is no space into which the TYR sidechain could move in the $5HT_{1D}$ to $5HT_{1B}$ mutation and therefore the selectivity area

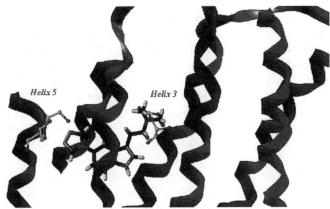

Figure 3 *The binding mode of a non-selective 5HT_{1D/1B} agonist which was the basis for the binding mode of the selective agonists*

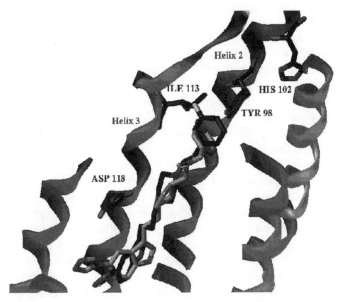

Figure 4 *An angled view of the selectivity determining region of the receptor with two selective compounds placed in their expected binding mode which illustrates how these compounds could interact with this region and how the amino acid sidechains themselves might interact*

becomes sterically congested which, in turn, affects the binding of the larger, selective compounds. With the reverse mutant the PHE is replaced by the smaller ILE residue which would allow the TYR much more freedom of movement. It is this freedom which, we believe, destabilises the receptor enough to weaken the binding of all compounds (this region being close to the key conserved ASP on TM 3). It was pleasing to see that the HIS102GLY mutation also affected the binding of selective compounds (affinities worsening by up to 12.5

fold), further implicating this region in the interaction with our selective compounds. The nature of this region is also sensitive to other changes as we have found that mutations of either TYR98, ASN105, TRP114 or SER117 produce a system which exhibited no radiolabel binding. It is interesting to note that the TRP114 residue is conserved, in terms of aromatic nature (*e.g.* PHE, TYR, ARG) across all 5HT receptors. This region is not totally intolerant of change though, as can be seen from the mutations of ILE100, GLN108, and ASP112 in the 5HT$_{1D}$ receptor which had no effect on the binding of our compounds.

Examination of the literature for mutations in this region reveals that only a small amount of work has been directed towards these residues. Of the 184 biogenic amine sequences listed in the Tromso GPCR mutation databases (GRAP and tinyGRAP in March 2001)[15,16] only 14 of them contain mutations within five amino acids of the important TYR/HIS residues on TM2 (three of these mutants are in fact chimeras) whilst only 15 sequences contain mutations within five amino acids of the ILE residue (six mutants being chimeras). However, some supporting evidence for the importance of this region has been seen in-house with the mutation of residues in the human dopamine D4 receptor to those of the D2 receptor. These mutants either reduce the binding affinity of the D4 selective ligands (*i.e.* the LEU112PHE mutation) or fail to produce a receptor which binds the radioligand (the MET113VAL mutation). It can be seen from the sequence alignment (Figure 1) that these dopamine residues fall very close to those of the 5HT$_{1D}$ receptor and their effects on selective compounds or receptor viability are comparable.

The change in binding affinity in going from human to dog 5HT$_{1D}$ receptors was unfortunate, as the dog is typically used as a safety assessment species. Many of the changes in the sequence occur in the third intracellular loop and onwards (see Figure 1). To determine whether these amino acids were responsible for the species differences, chimeric receptors were created at the LYS237 residue, combining the first five helices of the human receptor with the remainder from the dog receptor and *vice versa*. Interestingly no differences in the affinities of compounds for these chimeric receptors were seen compared to the wild-type receptors (data not shown). As the mutants around ILE113 had shown that they could influence the binding of the selective compounds we re-examined the human/dog sequence in that region. Of the three changes, we had previously mutated ILE100 and ASN105 in the human 5HT$_{1D}$/5HT$_{1B}$ receptors and found that there was either no effect, or no radiolabel binding respectively. The third difference in the sequences occurred at HIS102 which was now an ARG in the dog receptor. Mutation of the HIS to GLY in the previous experiments had shown an effect on our selective compounds and, gratifyingly, an effect (ranging from a 2.5- to 10-fold drop) on the binding of the selective compounds was seen when this residue was changed from HIS to ARG. The only literature support for our findings highlighted the key nature of this residue in the loss of binding affinity of 5HT$_{1D}$ receptor selective isochromans when moving from gorilla to guinea pig receptors.[17] Here the gorilla sequence has HIS102 whilst the guinea pig has ARG102 and most of the loss in affinity was regained when the mutant ARG102HIS was created. In contrast to our results where the ILE100 mutant produced no effect on compound affinity, they found that the THR100ILE mutant also improved the isochroman binding and the guinea pig double mutant returned the binding affinity to that seen in the gorilla receptor.

The other mutants to show some affinity changes were the replacement of PHE338 with ALA and the change of THR442 to VAL. As one can see from the image (Figure 5) the replacement of PHE338 with the smaller ALA residue should have enlarged the area around the protonated amine binding site, reduced the van der Waals (VDW) contact and therefore weakened the binding. This indeed was seen with the selective compounds losing 3–4-fold in affinity and the non-selective compounds only being slightly affected (1.5–2-fold change). The THR342 residue previously has been mutated to an ASN and shown to promote the binding of propanolamine compounds.[12,18] This was presumed to be due to a bidentate H-bonding interaction between the ether/hydroxyl oxygens of the propanolamine and the amide group of the ASN. It is interesting to note that the mutation of the THR342 into a VAL (*i.e.* removing all H-bonding possibilities) actual causes a 24-fold *increase* in the binding of a selective propanolamine derivative only. One presumes that the propanolamine is binding within the $5HT_{1D}$ receptor in a similar manner to other receptors (as it is the only compound to be greatly affected by the mutation) but its exact interaction with this residue cannot be dependent on H-bonding as it is in the other subtypes.

The recent publication of the 2.8 Å resolution crystal structure of rhodopsin,[19] the mammalian form of the receptor which provided the template for these homology models, led us to re-examine our models. However, as one can see from Figure 6, the homology model derived from the crystal structure does not allow a simple explanation of our mutagenesis data. This is due to the fact that whilst the helices are reasonably consistent, and hence the positions of TYR98 and HIS102, the loop region is quite different and as such ILE113 is situated some distance from its location in the bacteriorhodopsin model and directed away from the compound binding site. This displacement is due to the position of the second extracellular loop in rhodopsin which lies across the proposed

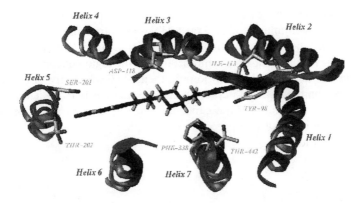

Figure 5 *View of the bacteriorhodopsin model with a selective compound placed in its expected binding mode. Some of the mutated residues are also shown*

Figure 6 *Overlay of the proposed bacteriorhodopsin based model (ribbon for the peptide backbone and sidechains in white) with a model based upon the recent rhodopsin crystal structure (ribbon and sidechain carbon atoms) and the rhodopsin crystal structure itself (tube for the peptide backbone). Illustrating how the differences between the two models could come solely from the orientation of the TM 4–TM 5 loop region*

binding area for our compounds, forming a 'lid' for the retinal binding, but also displacing the loop region at the top of helix 3. Considering the differences between the binding modes of retinal and the proposed biogenic amines, and also that we are dealing with agonists whilst the rhodopsin crystal structure is essentially in its 'dark' state, we believe that the rhodopsin crystal structure does not offer any quick improvements over our more simple bacteriorhodopsin based models.

7 Conclusions

This study provides further evidence for the existence of species differences in the pharmacology of the human, dog and rat 5-HT$_{1D}$ receptors and emphasises the need for caution when extrapolating data generated from animal to human tissue/receptors. We have shown the importance of the region encompassing the top of helices 2 and 3 and the loop joining them. The selectivity of our compounds depends on the critical interaction of three residues ILE113, TYR98 and HIS102, and also the viability of these biogenic amine receptors is dependent on the nature of a number of conserved residues within this region, *e.g.* TYR98, TRP114.

Acknowledgements

The authors would like to thank Dr. M. Russell, Mr. A. Reeve and Mr. G. Showell for their assistance with this paper.[20a–c]

References

1. A. Rabbeno, *Arch. Int. Pharmacodyn.*, 1949, **80**, 209.
2. F. G. Boess and I. L. Martin, *Neuropharmacology*, 1994, **33**, 275.
3. J. Longmore, D. Shaw, D. Smith, R. Hopkins, G. McAllister, J. D. Pickard, D. J. S. Sirinathsinghji, A. J. Butler and R. G. Hill, *Cephalagia*, 1997, **17**, 833.
4. P. R. Hartig, D. Hoyer, P.P.A. Humphrey and G. R. Martin, *TIPS*, 1996, **17**, 103.
5. T. A. Branchek, J. A. Bard, S. A. Kucharewicz, J. M. Zgombick, R. L. Weinshank and M. L. Cohen, *Experimental Headache Models*, Raven Press, New York, 1995, p.123.
6. J. M. Zgombick, J. A. Bard, S. A. Kucharewicz, D. A. Urquhart, R. L. Weinshank, and T. A. Branchek, *Neuropharmacology*, 1997, **36**, 513.
7. D. Oksenberg, S. A. Marsters, B. F. O'Dowd, H. Jin, S. Havlik, S. J. Peroutka, and A. Ashkenazi, *Nature*, 1992, **360**, 161.
8. QuikChange kits available from Stratagene Ltd, Cambridge Science Park, Milton Rd., Cambridge, CB4 4GF, UK.
9. QIAGEN DNA purification systems from QIAGEN GmbH Germany; see www.qiagen.com.
10. J. Sambrook and D. Russell, *Molecular Cloning: A Laboratory Manual*, Cold Spring Harbor Laboratory Press, USA, 2000.
11. N. Adham, P. Romanienko, P. Hartig, R. L. Weinshank and T. Branchek, *Mol. Pharmacol.* 1992, **41**, 1.
12. M. A. Metcalf, R. W. McGuffin and M. W. Hamblin, *Biochem. Pharmacol.*, 1992, **44**, 1917.
13. E. Parker, D. A. Grisel, L. G. Iben and R. A. Shapiro, *J. Neurochem.*, 1993, **60**, 380.
14. M. S. Beer, M. A. Heald, G. McAllister and J. A. Stanton, *Eur. J. Pharmacol.*, 1998, **360**, 117.
15. K. Kristiansen, S. G. Dahl and Ø. Edvardsen, *Proteins: Struct. Funct. Genet.* 1996, **26**, 81.
16. Ø. Edvardsen and K. Kristiansen, *7TM J.*, 1997, **6**, 1.
17. J. F.Pregenzer, G. L. Alberts, W. B. Im, J. L. Slightom, M. D. Ennis, R. L. Hoffman, N. B. Ghazal and R. E. TenBrink, *Br. J. Pharmacol.*, 1999, **127**, 468.
18. R. A. Glennon, M. Dukat, R. Westkaemper, A. M. Ismaiel, D. G. Izzarelli and E. M. Parker, *Mol. Pharmacol.*, 1996, **46**, 198.
19. K. Palczewski, T. Kumasaka, T. Hori, C. A. Behnke, H. Motoshima, B. A. Fox, I. L. Trong, D. C. Teller, T. Okada, R. E. Stenkamp, M. Yamamoto and M. Miyano, *Science*, 2000, **289**, 739.
20. (a) S. Bourrain, J. G. Neduvelil, M. S. Beer, J. A. Stanton, G. A. Showell and A. M. MacLeod, *Bioorg. Med. Chem. Lett.*, 1999, **9**, 3369; (b) A. M. MacLeod, L. Street, A. Reeve, R. A. Jelley, F. Sternfeld, M. S. Beer, J. A. Stanton, A. P. Watt, D. Rathbone and V. G. Matassa, *J. Med. Chem.*, 1997, **40**, 3501; (c) M. G. N. Russell, V. Matassa, R. R. Pengilley, M. B. van Niel, B. Sohal, A. P. Watt, L. Hitzel, M. S. Beer, J. A. Stanton, H. B. Broughton and J. L. Castro, *J. Med. Chem.*, 1999, **42**, 4981.

Computational Vaccine Design

Darren R. Flower*, Irini A. Doytchinova, Kelly Paine, Paul Taylor, Martin J. Blythe, Daniele Lamponi, Christianna Zygouri, PingPing Guan, Helen McSparron and Helen Kirkbride

EDWARD JENNER INSTITUTE FOR VACCINE RESEARCH
COMPTON, BERKSHIRE RG20 7NN, UK

1 Introduction

> 'That which does not kill us makes us strong'
> Friedrich Nietzsche, *Thus Spake Zarathustra*

Whatever subtle inferences Nietzsche may have wished his readers to draw from this maxim, he probably did not have an explicit reference to vaccinology in the forefront of his mind. However, this sententious and epigrammatic aphorism fits well the notion that we can do battle with the threat from infectious disease, and other dangers, by challenging our immune systems. These challenges may be artificial – vaccines, the topic of this chapter – or they may be naturally endemic or environmental in nature.

Of course, our immune systems face 'natural' challenges constantly. The so-called 'hygiene hypothesis' has suggested that in our urbanized, technologized, and increasingly comfortable, world, ostensibly beneficial improvements to personal hygiene and public health have, over decades, led to a widespread decrease in our exposure to pathogenic organisms. This has led to a decrease in the breadth and depth of acquired immunity to microbial pathogens and, in turn, to an increase in the prevalence of atopic disease: the increased tendency for individuals to make immediate, and inappropriate, hypersensitivity reactions to otherwise innocuous substances. The exposure to bacterial and viral pathogens early in life plays a significant role in the regulation of allergen-specific immune responses that underlie atopic allergy. Unless we accept that atopy is the price to be paid by certain populations for their freedom from microbial diseases, we must be prepared to continually train our immune systems, especially during infancy, in order to prevent allergic conditions. In centuries gone by, when hygiene was not manifest as widely or as well as it is today, this lack of challenge

---

* To whom correspondence should be addressed.

was not, of course, a great problem, but the direct threat from disease, albeit greatly exacerbated by poor diet and sanitation, was concomitantly greater. This can be seen most graphically in the devastating effects that old world diseases had on the population of the new world following the Spanish conquest in the early 16th century.

'Vaccine' is a term that can be applied to all agents, of either a molecular or a supramolecular nature, used to stimulate specific, protective immunity against pathogenic microbes, and the disease they cause, and ultimately to militate against the effects of subsequent infection. Vaccination, of course, pre-dates Nietzsche (1844–1890), beginning with the work of Edward Jenner (1749–1823). After a period of first training in London and then working for a time as an army surgeon, Jenner, a native of Gloucestershire, spent his entire career working in the county as a country doctor. On 14th May 1796, he used cowpox, a related virus, to build protective immunity against smallpox in his gardener's son. Later, Pasteur adopted 'Vaccination', the word Jenner had invented for his treatment (from the Latin *vacca*, a cow), for immunization against any disease. The influence of Jenner's work eventually led to the 1980 declaration by the World Health Organisation (WHO) that smallpox had been eradicated. It is now generally accepted that mass vaccination, taking account, as it does, of the principle of herd immunity, is one of the most effective approaches to the threat from infectious disease.

Although much effort is directed at disease treatment through the development of new antibiotic drugs, vaccines enjoy many intrinsic advantages. The frequency of vaccine treatment is, perhaps, in the range of once or twice per lifetime, to, say, once or twice per year. Compare this to the once or twice a day dosing required for most drugs. Moreover, it is possible to inject vaccines directly, thus circumventing many issues of bioavailability, which are increasingly complicating drug development. Vaccines are relatively cheap to produce if not necessarily to discover, making them of special interest to developing countries. On the down side of course, the development of new vaccines is not easy. Indeed, the development of vaccines has suffered from, amongst other things, the empirical nature of the vaccine discovery process. Many are still attenuated whole pathogen vaccines such as BCG, which is the vaccine in common use against tuberculosis (TB), and, perhaps, the most widely used vaccine in the world. In what follows, we will examine more modern, rational approaches to the design or discovery of vaccines, but we begin by first considering, in more detail, the threat from infectious disease.

1.1 The Threat from Infectious Disease

Infectious disease is one of most significant causes of death worldwide. It is well to remember, however, that it is only one of many, multifarious causes (see Table 1). Let us simplify them. In the Revelations of St John, the Four Horsemen of the Apocalypse were considered to be Pestilence, War, Famine, and Death. As an attempt to place the threat from disease into some kind of context, let us briefly examine each of these in turn.

Table 1 *Annual leading causes of death in the USA during 1990*

Cause of death	death rate[a]	Economic burden[b]
Heart disease	190	138
Cancer	135	104
Accidents	33	
Chronic pulmonary disease (*e.g.* asthma)	20	6
Pneumonia	14	
Infectious disease	12	
Diabetes mellitus	11	92
Suicide	11	
Homocide	10	
Chronic liver disease	9	

[a] Deaths per hundred thousand to the nearest whole number.
[b] Financial burden in billions of dollars to the nearest whole number.
Data compiled from figures in ref. 209.

Famine, the black horseman has, it may be argued, given rise to the greatest loss of life throughout history. Most quoted figures are probably great underestimates, as famine has always hit hardest the poorest and, to the retrospective eye of history, the most invisible. In modern times, for example, famines in Africa, most notably Ethiopia, and currently in Tajikistan have produced, and continue to produce, harrowing images of death and devastation. Looking back a little further to the famine of Northern China in 1969–71, history marks a death toll in excess of 20 million.

As the world population rises, the potential death toll from wars, and other conflicts, will rise with it. During the 20th century, the most populous in history, the total human population has risen from about 1.65 billion in 1901 to 6.08 billion in the year 2000. Although figures vary from between 40 million to 71 million, most historians agree that the total death toll of the Second World War was around 50 million, of which about 25% were military casualties. In comparison, the two largest death tolls from other international conflicts seem almost meagre: 15 million (First World War) and 2.8 million (Korean War). Compare those figures to loss of life through democide, or murder by government: around 48 million by the postwar China of Mao Tse Tung and the 20 million of Stalin's Russia. To the depredations of war we may add homicide, and other forms of violent death brought about by acts of volition. In the USA, for example, crude homicide rates are the third highest in the world: 4–73 times that in other industrialized nations. Between 1976 and 1993, more Americans were the victim of homicide than died on the battlefields of the Second World War. Homicide is the second leading cause of death amongst Americans aged 15–24, and the third leading cause among children aged 5–14. The legacy of war can persist long after the resolution of conflict: annually, for example, around 25,000 people are killed or seriously injured by land mines.

Death, the pale horseman, comes in many guises, covering diverse causes from individual natural disasters to accidental injury. Natural disasters, or what insurance brokers are pleased to call acts of god, would figure highly on the average individual's list of greatest causes of death and destruction. Floods, such as those seen in recent years in Mozambique, are an ever-present danger to life, livelihood, and property. A flood in the Henan province of China in 1939 caused the deaths of a million people. This was slightly more than a similar flood there in 1887, which cost the lives of 900,000. China is particularly prone to flooding: 1642 saw the deaths of three hundred thousand on the Huanghe river and a hundred thousand died on the Changjiang river in 1911. A typhoon caused the deaths of five hundred thousand in Bangladesh in 1970 and a tidal wave killed over 200,000 in the Bay of Bengal in 1876. Earthquakes have also caused death on a massive scale. For example, earthquakes in the Shanxi (1556), Tianjin (1976), Gansu (1920) and Hebei (1290) provinces of China killed an estimated 1.35 million between them. Likewise, volcanic action has resulted in large-scale death. The eruption of Krakatoa on August 27, 1883, though it did not kill anyone directly, resulted in a tsunami that killed over 36,000. The pre-historic eruption of Santorini was four times greater than that of Krakatoa and resulted in the destruction of the Minoan civilization. This was probably the greatest volcanic eruption in recorded history, though the probable eruption of Krakatoa during the sixth century, which caused worldwide climate change, was possibly larger. Volcanic action remains a threat, with over 70,000 people dying as its result during the 100 years of the 20th century.

At the other extreme, injury, however we may classify it, remains among the leading worldwide causes of death.[1] Injuries have, traditionally, been seen as primarily as random events, yet they affect all populations and act without reference to age, sex, income, or geography. In 1998, for example, about 5.8 million people, or a rate of, approximately, 100 per 100,000 population, died of their injuries throughout the world, causing around 16% of the global disease burden. Road traffic injuries are the 10th leading cause of death and the 9th leading cause of the burden of disease; falls and self-inflicted injuries follow closely.

Pestilence is often seen as the most terrifying of the four horsemen, although, in a world with increased awareness of potential bio-terrorism, some of these simplifying, artificial demarcations we have imposed are now becoming blurred. Disease is, however, also the one agent of human mortality that we can, in general, combat systematically through the use of biological and chemical entities, such as vaccines and drugs, through efforts of surgeons and physicians, and through improvements in public health, drinking water, and sanitation. Although it may be argued – and argued quite cogently – that the greatest benefit to man has come through improved public health, it is clear that vaccines and drugs have also made a huge contribution. In contrast, other than through their dispensing of drugs and other chemical therapies, the contribution made to public well-being by the trained medic, though the most direct, is relatively small.

Infectious disease is, then, the greatest source of preventable death. Both new diseases, of which the best known is human immunodeficiency virus (HIV, the cause of AIDS), and so-called re-emergent diseases, such as tuberculosis, figure equally amongst its list of causes. 5.3 million people became infected with HIV during the year 2000, bringing the total to around 36 million. Since the start of the epidemic over 21 million people have died from AIDS, of which over 3 million died during 2000. TB is the only infectious disease to be declared a 'global emergency' by the World Health Organisation. It is a chronic bacterial infection that causes more deaths worldwide than any other infectious disease. One-third of the world's population – around 1.7 billion people – are infected with the TB organism, *Mycobacterium tuberculosis*. Although most infected people never develop active TB, each year 8 million people do develop the disease across the world and 3 million die. The rapid spread of AIDS, especially in developing countries, has contributed to the sudden increase in TB cases in recent years. In fact, one third of the world's HIV positive population is now infected with the disease. Resistant strains of *M. tuberculosis* are also spreading for similar reasons.[2]

The need to thwart diseases such as TB has become imperative in an era of failing antibiotics. A recent report into antibiotic resistance by the World Health Organisation[3] detailed the most common resistant pathogens, and noted that formerly curable bacterial diseases were on the increase. For example, there has been a sharp rise in the number of nosocomial infections; up to 60% of these in the developed world are now caused by drug-resistant and often opportunistic pathogens like *Pseudomonas aeruginosa* and *Staphylococcus aureus*.

Several factors have contributed to the rise in resistance. In the five decades since penicillin became commercially available, misdiagnosed illness by health workers, patients failing to adhere to treatment, the widespread misuse of antibiotics with animals and the wrong prescription given for a particular disease, to name but a few, have all contributed to the problem. Within a competitive environment, resistance is able to spread quickly through the resident bacterial population. While antibiotics kill most susceptible cells, the residual resistant bacteria quickly colonize the empty niche, and pathogenic species can obtain a significant amount of their genetic diversity this way. Horizontal or 'lateral' gene transfer between distinctly related species adds to the problem;[4] resistance and/or virulence factors can be exchanged between a virulent donor and a recipient avirulent strain to produce new pathogenic varieties. Resistant *Neisseria gonorrhoea* now accounts for around 98% of all South-East Asia gonorrhoeal cases, and it is thought that the genes responsible for this arose from lateral transfer.

As antibiotic resistance increases, we may see history rewinding itself to the time of widespread plagues, epidemics, and pandemics. For most of the western world, this is an almost forgotten era, yet as recently as 1918 an influenza epidemic led to the death of over 22 million people worldwide. This is almost certainly significantly larger than the death toll from the First World War, yet visions of the trenches have all but effaced it from our collective memory. Perhaps the most destructive pandemic to ever have afflicted mankind was the Black Death. It accounted for over a third of Europe's population during the mid-14th century, costing the lives of approximately 75 million worldwide. The Black Death had first broken out in China in 1331 and began to spread westward. In October 1347, Genoese ships returning from the Crimea introduced the plague into Europe. The preceding forty years had laid a firm foundation for the success of the disease: poor harvests had led to a terrible Europe-wide famine in 1315; soon after, a typhoid epidemic killed thousands; in 1318 disease drastically reduced stock sizes; in 1321 another bad harvest brought more famine; and so on. By the end of 1350, the plague had spread through Europe as far as the North Baltic. Death was everywhere: we read of lawsuits where all parties died before their cases could be heard. We hear of dioceses where the surviving clergy were scarcely able to perform the last rites for their congregations. We are told of monasteries where half the inmates perished and of the Goldsmith's company of London, which had four masters within a year. The Black Death was a manifestation of bubonic plague, a systemic invasive disease, caused by the gram-negative bacterium *Yersinia pestis*. In the modern era, it has begun to show signs of multiple drug resistance. In all, it was responsible for three major human pandemics: the so-called Justinian plague (beginning in the spring of 542 and persisting till the 8th century), the Black Death (14th to 19th centuries, including the Great Plague of 1665) and modern plague (lasting from the 1850s till around 1960). There are, of course, many lesser epidemics of plague recorded before and since, starting with the first certain case, at least in the western world: the Libyan Plague of the early 1st century AD. Plague is one amongst very many epidemic diseases to have troubled humanity. For example, the Facts on File Encyclopedia of Plague and Pestilence lists over 600 separate, named, historically attested pandemics and epidemics from ancient times to the present.

One of the most significant events in the history of human disease interaction was the new world holocaust that affected South America in the century or so after its 'discovery' by the Spanish. The pre-Columbian inhabitants of the New World had the dubious distinction of having been isolated from the rest of humanity for longer than anyone else, and thus their exposure to pathogens which were endemic in the Old World. Exacerbated by the harsh treatment meted out by the Spanish, the Indian population fell victim to a number of diseases, foremost amongst which was smallpox. A few statistics will illustrate the point. When Columbus landed in Haiti in 1492, it had a population of 100,000; by 1570 it had fallen to 300. Peru's population of 1.3 million was halved in the fifty years between 1570 and 1620. In 1519, central Mexico had a population of 25 million – by 1605 this had dropped to around 1 million. The catastrophic decline in the indigenous Indian population was on a scale unmatched even in the 20th century, and was likely to have been the greatest ever loss of an aboriginal population.

The need, then, for new anti-microbial treatments, both therapeutic and prophylactic, is obvious. Antibiotic drugs are one approach to realizing such treatments. Vaccines are another. Historically, the most successful vaccination strategies have been based on attenuated whole pathogenic agents: viruses or bacteria. However, interest is now turning to more rationally designed vaccines. At one extreme of the size range, these can be genetically modified pathogens and at the other protein antigens or isolated T cell epitopes. Here the modern vaccinologist can make use of genomic data information from the many genomic sequencing projects focusing on pathogenic microorganisms. One of the key disciplines helping the discovery of new vaccines is the use of informatics strategies, primarily bioinformatics and molecular modelling. As we shall see in the rest of this review, these computational techniques are beginning to make a potentially important contribution to the next new generation of vaccine design programmes. We shall limit ourselves, in the main, to the discussion of vaccines directed towards cellular immune, or T cell, mediated responses.

2 Processing of Epitopes

A fundamental understanding of immunology underlies our attempts to design vaccines rationally. The manifestation of immunology at the whole animal level is, however, an exceedingly complex phenomenon, and it is only by investigating, at the molecular level, each of its individual stages, in a physico-chemical manner, that we are able to formulate effectively ways of modelling the process.

We shall concentrate our attention on that part of the adaptive immune response that is mediated by cells. A specialized type of immune cell mediates cellular immunity: the T cell, which constantly patrols the body searching out proteins that originate from a pathogenic organism, be that virus, bacterium, fungus, or parasite. The surface of T cells is enriched in a particular kind of receptor protein: the T cell receptor or TCR, which functions by binding to major histocompatibility complex proteins (MHCs) expressed on the surfaces of other cells. These proteins, in turn, bind small peptide fragments derived from both host and pathogen proteins. It is the recognition of such complexes that lies at the heart of the cellular immune response. These short peptides are referred to, by immunologists, as epitopes. The overall process leading to the presentation of antigen-derived epitopes on the surface of cells is a long, complicated, and not yet fully understood story. There are many alternative processing pathways, but we shall look at just the two major types: Class I and Class II (see Figure 1). Class I MHCs are expressed by almost all cells in the body. They are recognized by T cells whose surfaces are rich in CD8 co-receptor protein. Class II MHCs are only expressed on so-called 'professional antigen presenting cells' and are recognized by T cells whose surfaces are rich in CD4 co-receptors.

Class I peptides are ultimately derived from intracellular proteins, such as viruses. These proteins are targeted to the proteasome, which cuts them into short peptides of 8 to 11 amino acids in length. These peptides are then bound by

Figure 1 (opposite) *Immunological antigen presentation pathways.* (a) *Presentation of antigen via Class I MHC: Class I MHCs are expressed by almost all cells in the body. They are recognized by so called CD8+ T cells. Class I peptides are derived from intracellular foreign or self-proteins, which are targeted to the proteasome. This proteolytic macromolecular assembly cleaves them into peptides of length 8–11, which are then bound by the peptide transporter TAP. This translocates them from the cytosol into the endoplasmic reticulum where they are, in turn, bound by MHCs. The peptide–MHC complex is then exported to the surface of the cell through the golgi*

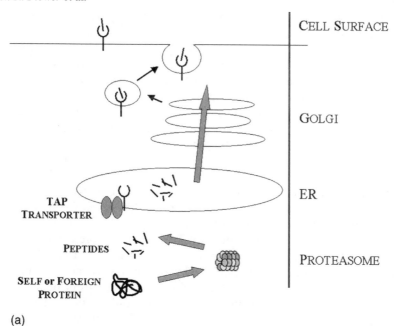

(a)

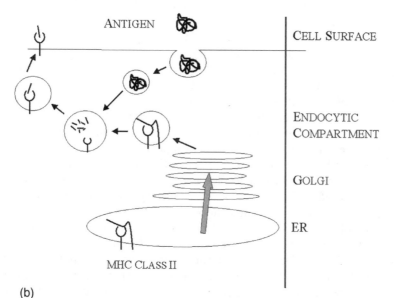

(b)

(b) *Presentation of antigen via Class II MHC: Class II MHCs are only expressed on 'professional antigen presenting cells' and are recognized by CD4+ T cells. In the Class II pathway, receptor mediated ingestion of extracellular antigenic protein is targeted to an endosomal compartment, where the proteins are cleaved by cathepsins, a particular class of protease, to produce peptides of length 15–20. Class II MHCs, exported through the golgi from the endoplasmic reticulum, then bind these peptides, displacing an endogenous peptide called CLIP, before transfer to the cell surface* the transmembrane peptide transporter TAP, which translocates them from the cell cytoplasm to the endoplasmic reticulum where they are bound by MHCs. Theoretical analyses of proteasomal cleavage patterns have been conducted by a number of groups,[5-7] leading in turn to a number of prediction methods,[8] some of which are available on the Internet.[9] The amount of data studied remains relatively small, and the predictive power the different methods have has yet to be evaluated objectively. Nonetheless, these represent useful contributions and important starting points for future study. Likewise, studies have also been conducted on the peptide substrate specificities of the TAP transporter,[10] leading to the development of predictive models[11] for the determination of peptides that bind to TAP. Together studies on proteasomal cleavage and TAP transport represent a good first attempt to produce useful, predictive tools for the processing aspect of Class I restricted epitope presentation. However, there are a number of other processing routes which complicate the simple picture outlined above. These include TAP-independent Trojan antigen presentation[12] and the involvement of various other proteases, such as furin.[13] Thus, the accurate prediction of epitope processing will need to rely on a much more comprehensive modelling of the entire process. This will account, perhaps through the use of mathematical modelling techniques prototyped on reaction kinetics within multi-enzyme metabolic pathways, as well as the bioinformatic modelling of cleavage patterns, for the complex hierarchy of interrelated dynamic processes that generate presented peptides.

For Class II, receptor mediated ingestion of extracellular protein derived from a pathogen is targeted to an endosomal compartment, where the proteins are cleaved by cathepsins, a particular class of protease, to produce slightly longer peptides of 15–20 amino acids. Class II MHCs then bind these peptides. The peptide specificity of protein cleavage by cathepsins has also been investigated and simple cleavage motifs are now well known.[14] However, more precise investigations are required before accurate predictive methods can be realized.

As we have said, peptide bound MHCs (or peptide–MHC (pMHC) complexes) are recognized by receptors on the surface of T cells, so called TCRs. Many other co-receptors and accessory molecules, in addition to CD4 and CD8 molecules, are also involved in T cell recognition. The recognition process is by no means simple, and remains poorly understood. Nonetheless, it has emerged that the process involves the formation of the so-called immunological synapse, a highly organized, spatio-temporal arrangement of receptors and accessory molecules of many types. The involvement of these accessory molecules, although essential, is not properly understood, at least from a quantitative perspective. Ultimately, the accurate modelling of all these complex processes will be required to gain full and complete insight into the process of epitope presentation.

3 Prediction of MHC Binding

The accurate prediction of T cell epitopes, much less immunodominant epitopes, remains problematic, and because of this, theoretical work, which is described below, has focussed on the prediction of peptide binding to MHC molecules. In order to quantify adequately the affinities of different MHCs for antigenic peptides, many different methods have been developed. It is possible to group these methods together thematically, based on the kind of underlying techniques they employ, and we shall endeavour to review them in this fashion below. As a preliminary, it is perhaps appropriate to mention some of the underlying issues involved.

Different MHC alleles, both Class I and Class II, have different peptide specificities. One way of looking at this is to say that they bind peptides with particular sequence patterns. This has led to the development of so-called motifs. A more accurate description of this phenomenon is to say that MHCs bind peptides with a binding constant dependent on the nature of the bound peptide's sequence. The driving forces behind this binding are precisely the same as those driving drug binding. Within the human population there are an enormous number of different, possible variant genes coding for MHC proteins, each exhibiting a different peptide-binding sequence selectivity. T cell receptors, in their turn, also exhibit different affinities for pMHC. The combination of MHC and TCR selectivities thus determines the power of peptide recognition in the immune system and thus the recognition of foreign proteins and pathogens.

Experimentally, there are many different ways of measuring binding affinity. IC_{50} values are binding affinity measures calculated from a competitive binding assay.[15–17] The value given is the concentration required for 50% inhibition of a standard labelled peptide by the test peptide. Therefore nominal binding affinity is inversely proportional to the IC_{50} value. Reference peptides can be labelled fluorescently or with a radioisotope. The results calculated from these two methods are significantly different, making their direct comparison difficult, and are therefore presented separately. BL_{50} values are calculated in a peptide binding stabilization assay.[18–20] It is the half maximal binding level calculated from a mean fluorescence intensity (M.F.I.) of MHC expressing RMA-S cells. These cells are incubated with the test peptide and then labelled with a fluorescent monoclonal antibody. The nominal binding strength is again inversely proportional to the BL_{50} value. The half-life for radioisotope labelled β_2-microglobulin disassociation from an MHC class I complex, as measured at 37 °C, is an alternative measure.[21–23] The greater the half-life the stronger the peptide-MHC complex. Apart from these three measures, many others are available. These include SC_{50}, C_{50}s, *etc.* which are closely related to BL_{50}s. Association and Dissociation equilibrium constants have also been measured, although far less frequently, and mean fluorescence intensities, measured at a single peptide concentration are, by contrast, very widely reported. These are amongst the most well reported measures, but there are many more. No clear consensus has, thus far, emerged on the most appropriate type of affinity measurement or assay strategy.

A widely used conceptual simplification, often used to help combine this bewildering set of binding measures, is to reclassify peptides as either non-binders or as high-binders, medium binders, and low binders. For example, the schema used by Brusic[24] classifies binders using these criteria: non-binders >

$10\,\mu M$, $10\mu M$ > low binders > $100\,nM$, $100\,nM$ > medium binders > $1\,nM$, high binders < $1\,nM$. Such broad schemes also allow for the inherent inaccuracy in MHC binding measurements.

Once a peptide has bound to a MHC to be recognized by the immune system, the pMHC complex has to be recognized by one of the TCR of the T cell repertoire. It is generally accepted that a peptide binding to an MHC may be recognized by a TCR if it binds better with a pIC_{50} > 6.3, or a halflife > 5 minutes, or some similar figures for other measures of other binding methods.[25] Some peptides binding at these affinities will become immunodominant epitopes, others weaker epitopes, and still others will show no T cell activity. There is some evidence suggesting that as the MHC binding affinity of a peptide rises, then the greater the probability that it will be a T cell epitope. The trick – the unsolved trick – is to determine which will be recognized by the TCR. Generally, the approach taken has been to whittle down the number of epitopes to a small number using prediction of MHC binding (see Figure 2). These peptides are then tested as potential epitopes in one of a great variety of different measures of T cell activation, such as T cell killing or thymidine incorporation, *inter alia*. The prediction, then, of MHC binding is both the best understood, and, probably, the

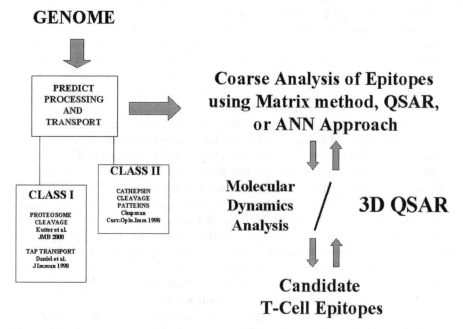

Figure 2 *Skeleton scheme for the prediction of T cell epitopes. A brief outline of the stages involved in the prediction of T cell epitopes. A genome, or, more properly, a proteome, is analysed from a presentation perspective (i.e. proteasomal cleavage or TAP) binding, with the potential peptides successively generated and excluded by this process, passing into a quick, but relatively crude, predictor of MHC binding which highlights potential binders. The resulting set of binders is then analysed by a more accurate, but more computationally demanding, method, such as molecular dynamics, to generate a list of candidate epitopes*

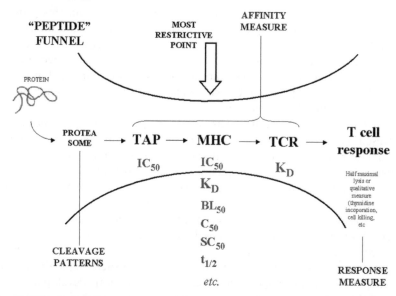

Figure 3 *Class I peptide presentation pathway. A schematic of the Class I presentation pathway showing stages in the processing of antigen to expressed epitope. The action of the proteasome has been characterized in terms of cleavage patterns. Binding of peptides to TAP, to MHCs, and the binding of PMHC to TCRs has been characterized in terms of affinity measures. Peptide binding to MHCs has, in particular, been characterized by a large collection of different quantitative measures. These can be divided into IC_{50}-like, BL_{50}-like, equilibrium constants of binding, and halflife. The response of T cells has also been characterized by a variety of measures, such as % cell killing or half-maximal lysis. These are whole cell or whole population measurements, rather than ones based solely on defined molecular interactions. It is thought that MHC binding is the most discriminating step on this presentation funnel*

most discriminating step in the presentation-recognition pathway (see Figure 3). In what follows, we will focus our attention on this stage.

3.1 Motif Approaches

The first attempts to computerize the identification of MHC binding peptides led to the development of motifs characterizing the peptide specificity of different MHC alleles. Such motifs – a concept with wide popularity amongst immunologists – characterize a short peptide in terms of dominant anchor positions with a strong preference for certain amino acids. Probably the first proper attempt to analyse MHC binding in terms of specific allele-dependent sequence motifs was by Sette et al.[26] They defined motifs for the mouse alleles I-Ad and I-Ed after measuring affinity for a large set of synthetic peptides originating from eukaryotic and prokaryotic organisms, as well as viruses; in addition they also assayed a set of overlapping peptides encompassing the entire staphylococcal nuclease molecule. Sette et al. quote prediction rates at the 75% level for these two alleles. A large number of succeeding papers, both from this group and others, have extended this approach to many other human and mouse alleles. Alleles from organisms other than these have not been studied in such detail, although the rat and bovine and some primate species have received some attention. An up-to-date compendium of such motifs is contained in the SYF-PEITHI database developed by Rammensee *et al.*[27] It is available on-line (http://syfpeithi.bmi-heidelberg.com/).

As we have said, these motifs are usually expressed in terms of anchor residues: the presence of certain amino acids at particular positions that are thought to be essential for binding. For example, human Class I allele HLA-A*0201, probably the best studied of all alleles, has anchor residues at peptide positions P2 and P9 for a nine amino acid peptide. At P2, acceptable amino acids would be L and M, and at the P9 anchor position would be amino acids V and L. Secondary anchors, residues that are favourable, but not essential for binding, can also be present. Moreover, sequencing of peptides that are known to bind shows preferences for particular amino acids at particular positions, although whether this represents anything other than the inherent bias in protein sequences is seldom addressed. Very many papers have been published both developing new, or refining old, motifs and also many papers that have tried to used them predictively. The method is admirably simple: it is easy to implement either by eye or more systematically using a computer to scan through protein sequences. Some computational approaches that use such motifs in a predictive fashion have extended standard sequence analysis methods to search for human class I peptides[28] and peptides that bind to bovine MHCs.[29]

However, there are many problems with the motif approach. Although it is possible to score the relative contributions of primary and secondary anchors to produce a rough and ready measure of binding affinity,[27,30] the most significant problem with the motif approach is that it is, fundamentally, a deterministic method. A peptide is either a binder or is not a binder. Even a brief reading of the immunological literature shows that matches to motifs produce many false positives, and are, in all probability, producing an equal number of false negatives, though peptides predicted to be non-binders are not always screened.

One useful outcome of work at this level, given the variety of different experimental techniques able to generate such motifs, is that it has clearly indicated that MHC alleles can be grouped together into so-called supertypes, which exhibit broad supermotifs, based on the commonality of their substrate specificity.[31] This may well prove useful in trying to rationalize more accurate prediction methods. Likewise, it can allow one to undertake more interesting analyses than the paucity of more sophisticated models will allow. For example, Zhang *et al.*[32] searched for Class I binding motifs in structural proteins of HIV of different genetic lineages to map the genetic organization of potential T cell antigenic sites. They found that local organization is characterized by clustering in relatively short regions, while large-scale organization shows segments of anomalous length between motifs. Low motif density occurs preferentially in variable portions of the protein sequence, suggesting that the virus may be generating escape mutants in order to evade T cell responses.

3.2 Experimental Matrices

While useful in themselves, binding motifs are, as we have said, very simplistic. They are not quantitative and their over-reliance on anchor positions can lead to unacceptable levels of false positives and false negatives. Alternative approaches abound. The different types have, as one might expect, different strengths and different weaknesses. The strategy adopted by many workers is to use data from binding experiments to generate matrices able to predict MHC binding. For want of a better term, we refer to these approaches as experimental matrix methods, as most such methods use their own measured data and relatively uncomplicated statistical treatments to produce their predictive models.

Reay *et al.*[33] for example, substituted all natural amino acids at the 11 positions of a moth cytochrome c epitope and evaluated binding to mouse Class II allele Iek. Apart from identifying three positions that significantly affected binding, they developed a simple scoring system to give semi-quantitative estimates of peptide-MHC affinity. Rothbard *et al.*[34,35] developed a method to predict the strength of binding to human Class II allele HLA DRB1*0401. They assumed, as do many other workers, that peptides of the same length bind similarly and that the contribution made by each side chain is independent and can be treated as a simple sum of residue interactions. Within the context of an otherwise polyalanine backbone the contributions made by the central 11 positions, of a 13 amino acid peptide, were quantified by measuring the effects of changes in amino acid identity at each position within the peptide. Later, this method was extended to further alleles: HLA DRB1*0101 and HLA DRB1*1501.[36]

An alternative strategy is the use of positional scanning peptide libraries (PSPLs) to generate such matrices. A number of such studies have been conducted. Some are aimed at investigating the problem of MHC-peptide interaction,[37-39] while others concern themselves with evaluating how variations in peptide sequence contribute to TCR recognition and T cell activation.[40,41] One of the most recent of these is also one of the most promising: Udaka *et al.*[42] have used PSPLs to investigate the influence of positional sequence variation on binding to the mouse Class I alleles Kb, Db, and Ld. From their analysis a program that could score MHC-peptide interaction was developed and used to predict the experimental binding of an independent test set. Their results showed a linear correlation but with substantial deviation. About 80% of peptides could be predicted within a log unit.

There are many other papers developing methods of this ilk. Though valuable contributions, it is clear that they betray a series of important limitations. Firstly, they do not, in general, constitute a systematic approach to solving the MHC–peptide binding problem. Rather, they are a set of different – essentially individual, independent, and inconsistent – solutions to the same, or nearly the same, problem. The measures of binding are different, the degree of quantitation is different for different methods and they also lack subsequent applications corroborating their predictive power. Moreover, few, if any, of the papers describing such work make available sufficient detail for others to use their methods independently, despite the relative simplicity of their respective computational approaches.

3.3 Empirical Methods

A step forward from deterministic motifs came with the work of Parker.[43] This method, which is based on regression analysis, gives quantitative predictions in terms of half-lives for the dissociation of β_2-microglobulin from the MHC complex. It is founded on a series of important observations about peptide binding to MHC molecules[21–23,44–50] and has been used in a number of applications.[51,52] Moreover, apart from its intrinsic utility, one of the other important contributions of this approach is that it was the first to be made available on-line (http://bimas.dcrt.nih.gov/molbio/hla_bind/). This method, often referred to as BIMAS by immunologists, is, for this reason, widely used. Because the underlying methodology was developed specifically to address immunology projects by Parker, rather than adopting an existing methodological approach, we choose to call it an empirical method, although this is, perhaps, a slightly inaccurate choice of terminology.

A number of other empirical methods also exist, each derived specifically with the prediction of MHC binding in mind. For example, Alves *et al.*[53] have used a combination of sequence alignments, phylogenetic dendrograms and calculated physical data to predict potential T cell epitopes from the cysteine proteinase of *Leishmania*. More recently, Borras-Cuesta *et al.* developed and compared algorithms for the prediction of Class II binding peptides.[54] These algorithms were based on matrices expressing the peptide selectivity of different alleles. The sensitivity and specificity of these algorithms were tested against different panels of peptides and compared to other algorithms reported in the literature. They note the rather unsurprising conclusion that the sensitivity and specificity of the approach was dependent on the prediction threshold. The sensitivities and specificities for test peptides were, however, similar to those used to develop their algorithms.

De Groot and colleagues have developed several computer programs, principlely EpiMer and EpiMatrix, and have used them in various practical applications, with a particular focus on HIV.[55–64] EpiMatrix and EpiMer are pattern-matching algorithms that attempt to identify putative MHC-restricted T cell epitopes as a preliminary to constructing multi-epitope vaccines. These algorithms are themselves based on matrix representations of positional amino acid preferences within MHC-bound peptides. The general utility of these methods has been limited by the commercial exploitation of the EpiMatrix and EpiMer technology.

Hammer and co-workers have developed an alternative computational strategy called TEPITOPE.[65–81] Good reviews of this methodology are now available.[82,83] Although the program can provide allele specific predictions, its main focus is on the identification of promiscuous Class II binding peptides. This has also been applied in a number of practical applications,[84–87] but, again, the general usefulness of the approach is limited by the commercial status of

TEPITOPE. One of the most interesting aspects of Hammer's work has been the development of so-called virtual matrices, which, in principle, provides an elegant solution to the problem of predicting binding preferences for alleles for which we do not have extant binding data.[88] Within the three-dimensional structure of MHC molecules, binding site pockets are shaped by clusters of polymorphic residues and thus have distinct characteristics in different alleles. Each pocket can be characterized by 'pocket profiles', a representation of all possible amino acid interactions within that pocket. A simplifying assumption is that pocket profiles are, essentially, independent of the rest of the binding site. A small database of profiles was sufficient to generate, in a combinatorial fashion, a large number of matrices representing the peptide specificity of different alleles. This concept has wide applicability and underlies, for example, attempts to use fold prediction methods to identify peptide selectivity. Other workers, such as Brusic, are developing similar technology.

3.4 Artificial Intelligence Methods

A number of groups have used techniques from artificial intelligence research, such as artificial neural networks (ANNs) and hidden Markov models (HMMs), to tackle the problem of predicting peptide–MHC affinity. ANNs and HMMs, are, for slightly different applications, the particular favourites when bioinformaticians look for tools to build predictive models. However, the development of ANNs is often complicated by several adjustable factors whose optimal values are seldom known initially. These can include, *inter alia*, the initial distribution of weights between neurons, the number of hidden neurons, the gradient of the neuron activation function, and the training tolerance. Other than chance effects, neural networks have, in their application, suffered from three kinds of limiting factor: overfitting, overtraining (or memorization), and interpretation. As new, more sophisticated neural network methods have been developed, and basic statistics applied to their use, overfitting and overtraining have been largely overcome. Interpretation, however, remains an intractable problem: few, if any, can easily visualize or interpret the very complex weighting schemes used by neural networks.

Notwithstanding these potential problems, many workers have adopted an ANN strategy in seeking to solve the prediction of peptide-MHC binding. Bisset and Fierz[89] were amongst the first to use ANN in this context. They trained an ANN to relate binding to the Class II allele HLA-DR1 to peptide structure and reported a correlation coefficient of 0.17 with a statistical significance of $p = 0.0001$. Adams and Koziol[90] used ANN to predict peptide binding to HLA-A*0201, which is probably the most abundantly assayed of all Class I alleles. They took a dataset of 552 nonamers and 486 decamers and generated a predictive hit rate of 0.78 for classifying peptides into two classes, one showing good or intermediate binding and another demonstrating weak or non-binding. Gulukota *et al.*[91,92] developed two complementary methods for predicting binding of 463 nonamer peptides to HLA-A*0201. One method used an ANN and the other used statistical parameter estimation. They found the ANN was better than motif methods for rejecting false positives, while their other alternative method was superior for eliminating false negatives.

Amongst the best known names of those interested in the area of MHC binding prediction is Vladimir Brusic. Over many years, he, and his co-workers, have developed a range of artificial intelligence techniques, including, *inter alia*, ANNs, HMMs, and evolutionary algorithms, aimed at solving problems of this kind.[93–100] His work contains models of both Class I and Class II MHC alleles, as well as the TAP transporter,[10,11] and, within the context of his own classification scheme,[24,101–105] his models seem highly predictive. Milik *et al.* used ANN to predict binding to the mouse Class I molecule Kb based on a training set of binding and non-binding peptides derived from a phage display library.[106] While it was easy to identify strongly affine peptides with a number of different methods, they found that ANNs predicted medium binding peptides better than simple statistical approaches.

Other varieties of artificial intelligence technique applied to this area include decision trees[107] and HMMs. Mamitsuka[108] has applied supervised learning to the problem of predicting MHC binding using an HMM as his inference engine. In a cross-validation test, the discrimination accuracy of their supervised learning method is usually approximately 2–15% better than other methods, including back propagation neural networks. Interestingly, his HMM model allowed the straightforward identification of new, non-natural peptide sequences that have a high probability of binding.

3.5 Quantitative Structure–Activity Relationship Methods

There are relatively few examples within the literature that apply Quantitative Structure–Activity Relationship (QSAR) methodology to questions arising from the immune system, nor indeed are there that many papers that apply QSAR techniques to any bioinformatic problem. The difference between QSAR and artificial intelligence methods is primarily a semantic one. In practice they achieve the same goal and work in similar ways, but QSAR techniques tend to be based on different, and possibly more rigorous, types of statistical analyses: including, amongst others, multiple linear and continuum regression, discriminant analysis, and partial least squares.

Ronna Mallios is one of the few long-standing exponents of this particular art.[109–112] Focussing on the problem of MHC Class II prediction, she has developed an iterative stepwise discriminant analysis meta-algorithm to derive a quantitative motif for classifying potential peptides as potential epitopes. Her most recent results, based in part on Brusic's MHCPEP database,[113] have allowed peptides to be classed as high-, moderate- or non-binders for HLA-DR1. Earlier work used Bayesian discriminant analysis to predict whether or not a given peptide epitope would activate helper T cells.[114]

In a recent paper, Chersi *et al*[115] used the MTD QSAR method to optimize binding of nonamer peptides to the MHC allele HLA-A*0201 by increasing the flexibility of amino acids at position 4. Bologa *et al.*,[116] again using the MTD approach, characterized the type of amino acid required for high binding of nine amino acid peptides at the HLA-A*0201 allele. They found that binding of these peptides is favoured by lipophilic side chains at positions 2 and 9 and by large amino acids at positions 1, 3 and 6. In a related study, Rovero *et al.*[117] analysed peptide binding to the human Class I allele HLA-B27. Most recently, Buss and co-workers have used statistical methods to develop and refine quantitative matrices representing Class I peptide specificities.[118] Their approach was to improve predictions by including sequence dependency. They developed an anchor-stratified calibration where their set of peptides was subdivided into groups containing peptides which had two, one, or zero primary anchors, leading to predictions with improved accuracy and precision.

In our own group we are beginning to apply a range of related QSAR techniques, including both 2D and 3D-QSAR methods, developed initially in pharmaceutical research, to attack similar MHC binding prediction problems. We have decided to adopt a quantitative approach, using widely available IC_{50} values, as measured in radioligand assays, rather than some arbitrary classification scheme. We have built this QSAR strategy on the foundation supplied by a new database system which we have developed.[119] This system, which we have called JenPep, is a group of relational databases that focuses upon quantitative data for peptide binding to MHCs and to the TAP peptide transporter, as well as an annotated list of T cell epitopes. The database, and a HTML graphical user interface (GUI) for its interrogation, is freely available *via* the Internet (http://www.jenner.ac.uk/JenPep).

The currently available version of JenPep (Version 1.0) is composed of three sub-databases: a compilation of quantitative affinity measures for peptides binding to Class I and Class II MHCs, a compendium of dominant and subdominant T cell epitopes, and a set of quantitative data for peptide binding to the TAP peptide transporter. The T cell section contains 2300 T cell epitopes, the MHC binding section contains 6000 peptides, and the TAP section covers 400 peptides. JenPep contains binding data on a wide variety of different MHC alleles: for MHC Class I, JenPep has data for 68 different restriction alleles with over 50 genotype variations. For Class II MHC molecules there are over 40 restriction alleles with 52 genotype designations. Peptide lengths for class I are in the range of 7–16 residues and for Class II are in the range of 9–35 residues. The database itself is a relational system, constructed using MicroSoft ACCESS and is searchable through a GUI built using ASP. Together with the peptide sequence, JenPep includes various kinds of binding measure, MHC restriction, and, where such data is known, the protein from which the peptide originates. Data on T cell epitopes is currently limited to a list of binders. While there are many different ways to identify T cell epitopes, including T cell killing, proliferation assays such as thymidine uptake, *etc.*, the quantitative data produced by such assays is not consistent enough to be used outside of particular experimental conditions. For MHC binding we have used a number of alternative measures of binding affinity which are currently in common currency. These include radiolabelled and fluorescent IC_{50} values, BL_{50} calculated in a peptide binding stabilization assay and kinetic half-lives.

We are actively developing the database beyond its current limitations, and expect to release a much larger and more complete quantitative database in the near future. Much useful data is still locked into the hard-copy literature or is presented in a graphical form, and it remains an on-going challenge to find and extract this data into a machine-readable format. We also look forward to the day when immunologists submit their experimental binding data to an online archive, much as today molecular biologists must submit their data to publicly curated sequence databases.

The approach we have taken to the prediction of MHC binding is to develop a novel QSAR application relating the biological activity of a set of peptides to their properties, as calculated in either 2D or 3D. The explanatory power of 3D-QSAR methods is considerable, not only in their ability to predict accurately biological activities, or, in our case, binding affinities, but also in their capacity to display advantageous and disadvantageous interaction potential, in three dimensions, mapped onto the three-dimensional structures of molecules being studied. The methods differ in the way they describe the compounds and in how they detect the relationship between 3D properties and bioactivity. We applied two well known 3D-QSAR techniques: Comparative Molecular Fields Analysis (CoMFA)[120] and Comparative Molecular Similarity Indices Analysis (CoMSIA).[121–123] These methods have some common features. They both use huge matrices of data generated at regularly spaced grid points using some distance dependent function. For the statistical calculation of the structure–activity relationship they both use a PLS protocol able to cope with the large volume of data. To obtain high predictive power they both need a perfect preliminary alignment of the investigated structures.

The main differences between CoMFA and CoMSIA are the form of the defined probe and the type of the distance dependent function used. In CoMFA, the probe atom is an sp^3 carbon with a $+1$ charge. For each molecule belonging to the set under study, two values of the interaction energy are calculated at each grid point – one a van der Waals/Lennard-Jones interaction and one an electrostatic Coulombic interaction. Because of the hyperbolic functional form, both of these potentials obtain very large nonsensical values within the van der Waals surface. To avoid these values arbitrarily fixed cutoffs are defined (here: $30\,kcal\,mol^{-1}$ for both functions). Due to the different slopes of the potentials these cutoffs are exceeded for the different terms at different distances from the molecules.[124]

In CoMSIA, similarity indices are calculated instead of interaction energies. Each molecule from the training set is compared to a common probe with a radius of 1 Å, charge, hydrophobicity, and hydrogen bond property equal to $+1$. The functional form here is selected to be Gaussian with an attenuation factor $\alpha = 0.3$.[121] Compared to the Lennard-Jones and Coulomb potentials, the Gaussian-type function has the advantage of using all grid points inside and outside the molecules, and no arbitrary cutoffs are required. Five different similarity fields are calculated: steric, electrostatic, hydrophobic, hydrogen bond donor and hydrogen bond acceptor. These fields cover the major contributions to ligand–protein binding.[122]

Initially, we applied two 3D-QSAR methods – CoMFA and CoMSIA – to a training set of 102 peptides that bound to HLA-A*0201.[125] The predictive power of both methods was assessed using a test set of 50 peptides. We found that CoMSIA gave much better predictive pIC_{50} values for binding to the HLA-A*0201 molecule than CoMFA, and indicated a dominant role for hydrophobic interactions in peptide binding to the MHC molecule.

More recently, we extended our 3D-QSAR analysis by applying the CoMSIA technique to a set of 266 peptides in order to assess the contributions of physicochemical properties other than the hydrophobic field.[126] The best model, based on steric, electrostatic, hydrophobic, hydrogen bond donor and hydrogen bond acceptor fields, had $q^2 = 0.683$ and $r^2 = 0.891$. The stability of this model was demonstrated by a 'leave-one-out' cross-validation procedure and by cross-validations in two and five groups. The mean |residual| value between the experimental pIC_{50}s ($-\log_{10}IC_{50}$) and those calculated by 'leave-one-out' cross-validation was 0.489. This model was used to evaluate the physicochemical requirements at each position in the peptide structure and to define the preferred amino acid sequence for high affinity binding to HLA-A*0201 molecule. The data are highly complementary to the sort of very detailed – but peptide specific – information obtained from crystal structures of individual peptide–MHC complexes.

We have also produced a complementary predictive method, based on the so called additivity concept. According to this concept each substituent makes an additive and constant contribution to the biological activity regardless of substituent variation in the rest of the molecule.[127] The IBS approach, as developed by Parker and others,[22,43,91] is based on a similar concept. We extended this idea by adding some additional terms accounting for the interaction of different side-chains. Thus, the binding affinity of a peptide will depend on the contributions of the amino acid side-chains at each position and the interactions between the adjacent and every second side-chain:

$$binding\ affinity = const + \sum_{i=1}^{9} P_i + \sum_{i=1}^{8} P_i P_{i+1} + \sum_{i=1}^{7} P_i P_{i+2}, \quad (1)$$

where the *const* accounts, albeit nominally, for the peptide backbone contribution, $\sum_{i=1}^{9} P_i$ is the sum of amino acids contributions at each position, $\sum_{i=1}^{8} P_i P_{i+1}$ is the sum of adjacent peptide side-chain interactions, and $\sum_{i=1}^{7} P_i P_{i+2}$ is the sum of every second side-chain interactions.

We applied this method to a set of 340 nonamer peptides with affinity to HLA-A*0201 molecule.[128] The regression equation consisted of 1815 terms including the constant. Its MLR parameters were $r^2 = 0.898$, $s = 0.285$, $F_{5,334} = 588.883$, number of components (NC) = 5. The 'leave-one-out' cross-validation (CV-LOO) gives $q^2 = 0.337$, SEP = 0.726, NC = 5, mean |residual| value = 0.573. The combination of both techniques – CoMSIA and additive – gives a very useful result. CoMSIA can make extrapolations, predicting the binding affinity of a peptide carrying an amino acid not present in the investigated set at the same position, but it cannot assess the contribution of each amino acid at each position and the interactions between them. The opposite is true for the additive method, it lacks the ability to extrapolate but it can give a quantitative assessment of any amino acid at any position in the peptide. The application of both methods to the same set of nonamers binding to MHC class I molecule HLA-A*0201 gave very good agreement between the results generated by both techniques.[129]

3.6 Structural Analyses: Roadmaps of MHC Binding

An alternative approach to gaining some kind of structural understanding of peptide binding proceeds through a thorough analysis of MHC receptor binding site structure. This is in contrast with data driven models that rely on the accumulation of binding data. Studies of that kind provide important structural information that underlies binding and it also allows links to be drawn between different MHC alleles, but at the functional, or peptide binding, level rather than the phylogenetic level. MHCs are polymorphic and it is small variations in the amino acid identity of binding site residues that give rise to variations in peptide selectivity exhibited during peptide binding. Thus any significant propinquity apparent between binding sites should also be mirrored in the overall peptide selectivities of different MHCs. Rigorous comparative investigation of similarities should allow us, then, to predict both similarities in peptide selectivities and to group different alleles together on a fundamentally different basis to those produced by pseudo-phylogenetic studies of the whole MHC sequence.

We have already mentioned this concept above, in the context of Hammer's work,[88] for example. A number of early studies of this type focussed on the then unclear nature of the directionality of peptide binding to MHC molecules.[130,131] More recent work has concentrated more on the nature of peptide binding and the kinds of sites and sub-sites apparent in MHC molecules, particularly in the context of Class I. De Lisi and co-workers,[132] sought to establish correlations between Class I peptide binding specificities and MHC sequence markers occurring at polymorphic positions within the binding site. The analysis of nine MHC crystal structures of Class I MHC molecules, together with the modelled structures of 39 more, suggested Class I pockets can be classified into families distinguishable by their common properties. In turn, this allowed the set of known Class I motifs to be greatly expanded. Sacchetini and colleagues again analysed human and mouse Class I molecules,[133] attempting to identify non-covalent interactions – hydrogen bond and van der Waals interactions – conserved between different structures. These interactions are characteristic of individual MHC Class I molecules, and determine the nature of anchor residues in bound peptides.

The name of Chelvanayagam stands out amongst work of this type. He has attempted to derive so-called 'road maps' for both Class I[134,135] and Class II[136,137] MHC molecules. In Chelvanayagam's work, new, less restrictive descriptions of the peptide binding sites of MHC molecules are developed. Chelvanayagam refers to these as peptide binding environments and defines them as that set of amino acid residues within a preset neighbourhood of individual residues in crystal structures of MHC peptide complexes. Combining this information with sequence alignments of Class I MHCs, Chelvanayagam is able to make predictions for motifs for those MHC molecules that share a similar profile of environments. Chelvanayagam has extended this approach seeking to rationalize the six supertypes of MHC HLA-A molecules defined on the basis of nucleotide sequence and phylogenetic analysis.[138] Unsurprisingly, he finds closer approximation to family signatures defined at sites showing strong correlation with these six groups. In particular, positions 62, 97 and 114, within the multiple alignment, can discriminate between these families. Chelvanayagam finds that while the whole site contributes to the definition of antigen binding, these three amino acid positions play the most important role in the determination of supertype specificity and the nature of T cell recognition.

In an interesting variant on this kind of static analysis, Schueler-Furman and co-workers have studied possible structural preferences of MHC-binding peptides by examining the conformation space defined by the structures of these peptides within the proteins from which they derived.[139] Comparing the conformational space accessible to a set of nonameric MHC binders and set of random nonamers showed no significant difference. This suggests that the MHC binding site has evolved to bind peptides with any 'structural background'.

3.7 Molecular Dynamics

A quite different approach to obtaining predictions of peptide–MHC binding is based on atomistic molecular dynamic simulations. It attempts to calculate the free energy of binding for a given molecular system, which is closely related to experimentally observable quantities such as equilibrium constants or IC_{50}s. It has the advantage that, in principle, there is no reliance on known binding data, as it attempts the *de novo* prediction of all relevant parameters given certain knowledge of the system. Essentially, all that is required is the experimentally determined structure, or a convincing homology model, of a MHC peptide complex.

The statistical mechanics definition of free energy is in terms of the partition function but this theoretical definition is not practical for most types of calculation. What one can calculate more easily, however, is the free energy difference between two states. A number of methods exist which allow us to undertake simulations leading to the evaluation of free energies, each based on different assumptions and offering differing levels of approximation. In free energy perturbation, or FEP, methodology, for example, the free energy is calculated at discrete intervals j using the expressions:

$$G_{\lambda(j+1)} - G_{\lambda(j)} = -RT \ln <e^{V_{\lambda(j+1)} - V_{\lambda(j)}}>_{(j+1)}$$

$$\Delta G = G_1 - G_o = \sum_j G_{\lambda(j+1)} - G_{\lambda(j)}$$

where G_1 and G_0 are the free energies of the two states and V is the potential energy of the system. In thermodynamic integration, the free energy of the system is calculated by:

$$\Delta G = G_1 - G_o = \int_0^1 d\lambda \langle dV/d\lambda \rangle$$

To solve this equation numerically we must transform it, from a continuous integration over a continuum of indivisible steps, to a discrete integration over a set of individual steps. In slow growth methods, we assume the steps are very close and we can approximate:

$$\frac{\delta H}{\delta \lambda} \sim \frac{\Delta H}{\Delta \lambda}$$

$$\Delta G = H_{\lambda = o} - H_{\lambda = 1}$$

Many other approximations also exist, notably, that introduced by Aqvist.[140]

Molecular dynamics simulation is, itself, a technique to compute the equilibrium position of a classical multiple body system. It is assumed that the atoms of the system are constrained by an interatomic potential energy force field. Each of the N atoms in the simulation are treated as a point mass and Newton's equations are integrated to compute their motion. This can be written in the formalism of Hamiltonian mechanics as:

$$x = JdH(x)$$

where J is the identity matrix of rank 2 and $H = T + U$, where T is the kinetic energy and U is the potential energy. We need to provide the initial configuration of the system at $t = 0$, that is the co-ordinates of all atoms in a six dimensional hyperspace. Thus, at regular time intervals, we resolve the classical equation of motion represented by the N equations implicit above. The gradient of the potential energy function is used to calculate the forces on the atoms while the initial velocities on the atoms are generated randomly. At this point new positions and velocities are computed and the atoms moved to these new positions. To measure an observable quantity we must be able to express this as the position in a phase space of dimension $6*N$. The information within our system is largely contained within the potential energy function, which takes the form of a simple penalty function for most simulations of biomolecules.

For large molecular systems comprising thousands of atoms, many of the more sophisticated modelling techniques, which often describe the potential energy surface in terms of quantum mechanics, are too demanding of computer resources to be useful. The Born–Oppenheimer approximation states that the Schrödinger equation for a molecule can be separated into a part describing the motions of the electrons and a part describing the motions of the nuclei and that these two motions can be studied independently. We can then think of molecules as mechanical assemblies made up of simple elements like balls (atoms), rods or sticks (bonds), and flexible joints (bond angles and torsion angles). Terms that describe the van der Waals, electrostatic, and possibly hydrogen bonding, interactions between atoms supplement molecular mechanics forcefields.

Delisi and co-workers were among the first to apply molecular dynamics to peptide:MHC binding, and have, subsequently, developed a series of different methods.[141-145] Part of this work has concentrated on accurate docking using molecular dynamics and part on determining free energies from peptide MHC complexes. Rognan has, over a long period, also made important contributions to this area.[146-152] In his work, dynamic properties of the solvated protein–peptide complexes, such as atomic fluctuations, solvent accessible surface areas, and hydrogen bonding patterns correlated well with available binding data. He has been able to discriminate between binders that remain tightly anchored to the MHC molecule from non-binders that are significantly weaker. Other work from Rognan and co-workers[153,154] has concentrated on the design of non-natural ligands for MHC molecules, demonstrating the generality of molecular dynamic approaches to problems of MHC binding.

Other work in this area has come from two directions. First, those interested in using the methodology to analyse and predict features of peptide–MHC complexes. These methods have looked at both Class I[155,156] and Class II,[157] as well as investigating the effect of peptide identity on the dynamics of T cell interaction.[158] Secondly, those who are more interested in developing novel aspects of MD methodology, including both simulation methodology[159] and solvation,[160] and use the MHC peptide systems as a convenient example of a binary molecular complex.

The growth of computer power during the last two decades has allowed the study of biologically interesting systems including small and medium-sized proteins using atomistic molecular dynamics methodology. However, we are still faced with problems concerning the validity of our models and the relatively short time scales that can be reached on current serial machines. Many approaches have been tried to circumvent these problems, but only with limited success, since almost any attempt to reach longer time scales will result in more approximations in the model. Previous attempts to utilize molecular dynamics and other atomistic simulation methods to investigate peptide–MHC interactions have foundered on technical limitations within present computing methods. While many methods link thermodynamic properties to simulations, they take an unrealistically long time. A basic simulation yielding a free energy of binding requires something like 10 nanoseconds of simulation. On the average desktop serial workstation, this requires a compute time in the order of 300 hours per nanosecond. To simulate as few as a dozen peptides might occupy a whole machine for several years. To circumvent these technical limitations one recourse we might make use of is to take advantage of high performance, massively parallel implementations of molecular dynamic (MD) codes running on large supercomputers with 128, 256, or 512 nodes. Another, complementary way to circumvent this problem is to make use of 'Grid computing'. This refers to an ambitious and exciting global effort to develop an environment in which individual users can access computers, databases and experimental facilities simply and transparently, without having to consider where those facilities are located. It is named by analogy with the national power transmission grid. If one wants to switch on a light or run a fridge freezer, one does not have to wait while current is downloaded first, thus Grid seeks to make available all necessary compute power at the point of need.

Mention should also be made of the study by Schueler-Furman and co-workers,[161] which strikes a cautionary note. A preliminary to the accurate prediction of binding data for MHC complexes is the fast, accurate modelling of the initial peptide structure in the MHC-binding groove. Using 23 Class I peptide–MHC complexes solved experimentally as a reference, Schueler-Furman evaluated algorithms for this purpose. The peptide backbone and MHC structures were used as rigid templates within which sidechain conformations were built from a rotamer library using the 'dead end elimination' method.[162] Within this context, they evaluated the influence of different parameters on the prediction quality. They concluded that the structure of the peptide sidechains is dictated by the structure of the static MHC molecule and that the prediction of individual rotameric states for specific amino acids is not affected significantly by sidechain–sidechain interactions. However, under cross validation the success rate in correctly predicting sidechain rotamers did not exceed 70%, indicating a fundamental limitation of existing modelling technologies.

3.8 Virtual Screening

A methodology closely related to molecular dynamics, both being based, to a large degree, on molecular mechanics force fields, or, at least, drawing on analogies from pairwise atomistic potential energy functions, is a set of techniques grouped loosely under the name of 'Virtual Screening'. There are two principal types of virtual screening methodology that have, thus far, been applied to the prediction of MHC binding. One derives from computational chemistry and the other from structural bioinformatics and the development of tools for fold prediction. Virtual screening is an expression deriving from pharmaceutical research: the use of predicted ligand–receptor interactions to rank or filter molecules as an alternative to high throughput screening. Approaches to virtual screening cover a spectrum of methods which vary in complexity from molecular descriptors and QSAR variables, through simple scoring functions (such as Ludi, FlexX, Gold, or Dock), potentials of mean force (PMF) (such as Bleep), force field methods, QM/MM, linear response methods,[140] to free energy perturbations. In this transition from, say, atom counts, through to full molecular dynamics, we see a tremendous increase in computer time required. Virtual screening can be seen as seeking a pragmatic solution to the accuracy gained *vs.* time taken equation. The point at which one stops on this spectrum is contingent upon the system being evaluated, the number of peptides being evaluated, and the computing resources available.

Rognan has developed a virtual screening method called FRESNO and applied this algorithm, which relies on a simple physicochemical model of host-guest interaction, to the prediction of peptide binding to MHCs.[163] This model was trained on a combination of data and experimentally derived 3D structures from the alleles HLA-A0201 and H-2Kk. They found that lipophilic interactions contributed the most to HLA-A0201-peptide interactions, whereas H-bonding predominated in H-2Kk recognition. Cross-validated models were afterward used to predict the binding affinity of a test set of 26 peptides to

HLA-A0204 (an allele closely related to HLA-A0201) and of a series of 16 peptides to H–2Kk. They concluded from their initial study that their scoring function was able to predict, with reasonable accuracy, binding free energies from three-dimensional models. In a more comparative study,[164] Rognan and colleagues found that, for predicting the binding affinity of 26 peptides to the Class I MHC molecule HLA-B*2705, FRESNO out-performed six other available methods (Chemscore, Dock, FlexX, Gold, Pmf, and Score). This confirms our own experience using commercial and freeware virtual screening approaches for the quantitative assessment of MHC peptide binding. Kanguene and colleagues[165] obtained a 77% success rate using the number of clashes between the MHC and peptide and the number of exposed hydrophobic peptide residues to correctly distinguish peptides into binders and non-binders.

Turning now to bioinformatic based approaches, others are using amino acid pair potentials, initially developed to predict the fold of a protein, to identify those peptides which will bind well to a MHC. Margalit and colleagues have proposed a number of virtual screening methodologies,[166,167] each of increasing complexity. They used amino acid pair potentials, originally developed by Miyazawa and Jernigan,[168] to evaluate the inter-protein contact complementarity between peptide sequences and MHC binding site residues. They presented an analysis of peptide binding to four MHC alleles (HLA-A2, HLA-A68, HLA-B27 and H–2Kb), and were successful in predicting peptide binding to MHC molecules with hydrophobic binding pockets but not when MHC molecules with charged or hydrophilic pockets were investigated. Again focusing on Class I alleles, a more recent study from this group[169] used an updated set of statistical pairwise potentials. These were developed from the Miyazawa and Jernigan potential by Betancourt and Thirumalai[170] and describe the hydrophilic interactions more appropriately. This enables more accurate modelling of the threading of the candidate peptide sequence. In an independent study, Swain et al.[171] have developed a similar threading method and applied it to Class II MHC–peptide interactions. This method is currently being developed commercially by Biovation Ltd [http://www.biovation.co.uk/].

Because of the relative celerity of virtual screening methods compared with MD methods and its ability to tackle MHC alleles for which no known binding data is available, this method has huge potential. While both molecular dynamics and related methods hold out the greatest hope for true *de novo* predictions of MHC binding, their present success rate is very much lower than that of data driven models. However, as with most of science, one must tease the genuinely useful from the self-aggrandizing hyperbole. Much work remains to be done on developing, refining, and applying this methodology.

4 Predicting T cell Epitopes

TCRs are immunoglobulins, homologous in sequence and structure to antibodies, and function through binding to the peptide–MHC (pMHC) complex formed on the surface of other cells. The binding of TCRs to pMHC is weaker (in the micromolar range) than the binding of peptide to MHC (in the nanomolar range). T cell epitopes are, as defined above, short peptides which bind well to either Class I or Class II MHC molecules and are recognized by TCRs, and thus activate a diversity of T cell responses. A number of workers have attempted to avoid the necessity of predicting MHC binding and opted instead for a direct evaluation of the potential of a peptide to become a T cell epitope.

Lu *et al.*[172] found, in sequences of digested fragments of antigenic proteins, a pattern of recurring hydrophobic sidechains forming a longitudinal hydrophobic strip (assuming an α-helical conformation) associated with T cell epitopes. Raychaudhury *et al.*[173] used a graph-theoretical method to extract allele specific patterns characteristic of T cell epitopes. In cross validated tests, they found their algorithm, which used indices calculated from weighted connected graphs to model their peptides, was almost 100% predictive. However, their database of only 28 peptides was so small that the statistical significance and global predictivity of the method must be questionable. Davenport *et al.*[174] used simple two or three anchor position motifs to predict MHC Class II restricted immunodominant T cell epitopes. Hobohm and Meyerhans[175] developed an automated procedure to extract anchor residue motifs from sets of Class I MHC binding peptide sequences: their motifs for A*0201, B27, Kb, Kd, Db were in substantial agreement with measured data. Their method was then used to predict the natural short epitope inside longer antigenic peptides. Altuvia *et al.*[176] developed a method which seeks to discriminate between true T cell epitopes and other non-epitopes, which are peptides that are inactive either because they are not recognized by TCRs or are not bound by MHCs. Again, they used multiple sequence alignments to generate motifs that are present in epitopes and absent in non-epitopes. A motif is expressed in terms of physico-chemical and structural properties that may give rise to the sequence specificity of binding and can be extracted from sequence data, such as hydrogen bonding capability, hydrophobicity and charge. The effectiveness of these motifs was explored using mouse Class II alleles. In a novel application of artificial intelligence techniques, Savoie *et al.*[107] used decision trees, as implemented in the BONSAI program, to identify sequence motifs that cause preferential activation of T cells. By using a database of previously identified T cell epitopes they were able to identify motifs that explained 84% of negative T cell responses and 70% of positive T cell responses.

In the light of X-ray crystallographic data, it is now clear what the mechanism of pMHC–TCR interaction is. However, this has not always been clear.[177] Early analysis of experimental results did show that T cell epitopes could be defined by short linear sequences, but the observation that increasing peptide length often resulted in increased antigenic potency suggested that immunogenicity might also depend upon the ability of the peptide to adopt an appropriate secondary structure. These ideas led many researchers to develop direct predictors of T cell immunogenicity based, in the main, on such erroneous assumptions about the conformation adopted by T cell epitopes. In an important retrospective analysis of these diverse approaches, Deavin *et al.*[178] compared numerous T cell epitope prediction methods against databases of mouse and human epitopes, assessing their performance using specificity as a measure of the quality of predictions and sensitivity as a measure of the quantity of correct predictions. *Versus* the human data, the strip-of-helix algorithm of Stille *et al.*[179] was the only significant model and for the mouse dataset, only the method of Rothbard and Taylor[180] was significant. Overall, they found that most of the algorithms were no better than random for either data set, indicating again the need to include MHC binding in the definition of putative T cell epitopes. This is consistent with the diminishing interest in such methods relative to that which predicts MHC binding. One of the few approaches of this type that has been published in recent years is the report by Cornette *et al.*,[181] which performed a Fourier analysis of sequences from a compendium of T cell epitopes. This indicated that peptides exhibited a periodic variation in amino acid polarities of 3–3.6 residues per period, suggestive of an amphipathic α-helical structure, which is clearly in conflict with the results of high resolution structural studies. They offered two explanations for this inconsistency based on the spacing of hydrophobic and hydrophilic residues and suggested the existence of an allele independent antigenic motif inherent within the structure of most T cell epitopes.

5 Predicting Subcellular Location

There are obviously many aspects to computational vaccine design other than the prediction of potential epitopes. Many of them are as yet only poorly developed. While we have seen that T cell epitope prediction is now well developed, at least to the stage where it is beginning to become useful, the prediction of immunogenicity, particularly for subunit vaccines, which necessarily involves a deeper understanding of host responses, remains primitive (see Figure 4). The prediction of antibody, or B cell, mediated antigenicity is at an even more primitive stage.[182,183] This relies on concepts of some antiquity[184–186] and quite simplistic software.[187,188] However, some other techniques complementary to the prediction of host responses, such as the prediction of the subcellular location of potential antigen proteins, have reached a greater level of maturity.

Consider a microbial genome, or, more specifically, a bacterial genome (Figure 5). The total protein complement – say a few thousand gene-products – is distributed between the inner and outer compartments of the bacteria. Some will reside in the cytoplasm, some will find their way to the periplasmic space, at least in Gram −ve bacteria, and others will be secreted from the cell. Some proteins will become integral membrane proteins located in the inner or outer membranes and some will become lipoproteins. An ability to predict these locations would be of great benefit when choosing which proteins to investigate as candidate vaccines: a secreted protein, for example, can be regarded, at least naively, to be a more likely target than, say, a cytoplasmic enzyme. A number of bioinformatic methods have been developed which address the prediction of subcellular location, which has proved to be more complex than was originally envisaged.

In 1982 a strong link between amino acid composition (*e.g.* Leu and Trp favoured, Pro disfavoured[189]) and cellular location was identified,[190] but as the

BIOINFORMATIC VACCINE DESIGN

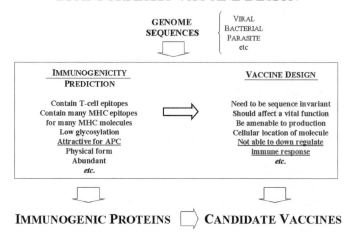

Figure 4 *Skeleton scheme for predicting immunogenic protein molecules and candidate subunit vaccines. The identification of protein immunogenicity and the identification of potential vaccine targets from a given genome, or set of proteins, is, as yet, an unsolved problem. The scheme given here uses the prediction of T cell epitopes to effect the leverage of a partial solution for the prediction of immunogenicity. However, many other factors, including many not cited here, will also affect it. Assuming that immunogencity can be predicted, then other factors can be used to filter highly immunogenic proteins as putative vaccine targets. Many of the factors shown here are tractable from a bioinformatic viewpoint, albeit with greatly varying degrees of accuracy, insofar as data exists from which we can build models, but those shown underlined require other novel experimental evidence. One can imagine a viral genome of a few hundred proteins, a bacterial genome of a few thousand, or a parasite genome of, say, ten thousand genes being fed into a composite software package that undertakes such an analysis, producing tens of highly immunogenic proteins, from which it can be whittled down to say five candidate vaccines*

number of protein structures available increased this relationship has became more blurred.[191] Despite the ambiguous relationship between amino acid composition and subcellular localization many methods, of increasing sophistication, have been created that exploit this connection.[192–194] Nakashima and Nishikawa[195] describe a method where the average amino acid composition for a number of proteins whose subcellular localization is known was calculated. From these simply obtained results trends in amino acid composition were observed such as intracellular proteins being relatively rich in aliphatic residues. Just using basic rules like these they were able to correctly identify 78% of the test set as being either intracellular or extracellular.

This idea was developed further by Andrade *et al.*[196] who hypothesized that throughout evolution each subcellular location has maintained a characteristic physio-chemical environment. The proteins in each location would have adapted to the environment and therefore each location would have proteins with signature structural characteristics. These characteristics are more likely to manifest at the surface (which is exposed to the environment) and therefore the surface

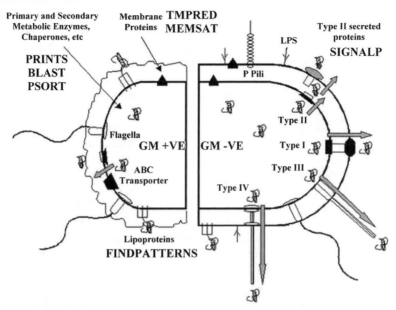

Figure 5 *Determining the subcellular location of proteins in Gram +VE amd Gram −VE Bacteria. Many methods exist for determining the subcellular location of proteins within the expressed set of gene products within a bacterium. Intracellular proteins, which are typically, but not exclusively, well conserved, can be identified through the use of global similarity homology searches of sequence databases, using BLAST for example, or motif databases, such as PRINTS, or through the use of generic subcellular location methods such as PSORT. Membrane proteins are readily, and quite reliably, found through the use of many different prediction methods. Lipoproteins can be found through the use of sequence patterns, and proteins secreted through the type II secretion system can be detected using SIGNALP. Proteins secreted through the other secretion systems I, III, IV are hard to detect, as they have no readily detectable secretion signals within their sequence*

residue composition is likely to give a very strong identification of the subcellular location. This method predicted 77% of protein locations accurately. Although amino acid composition correlated with subcellular location the former can not be exclusively defined by the latter. Neural networks have also been applied to this problem[197] and are the basis of the NNPSL web-based server. This provided an accuracy of 81% for prokaryotic prediction but only 66% for eukaryotic. This seems likely to be due to the persistent neural network shortcoming of overfitting to training data especially when the variables are complex.

The majority of methods for predicting localization are based on protein sorting signals.[198] These signals are normally represented as a short sequence with variable levels of conservation. Many are represented as well defined motifs while others show vague sequence features that are undetectable by simple homology searching.[199] The most obvious protein sorting signal to investigate is the signal peptide. Looking at a simple bacterial model, if a protein has a signal peptide but no transmembrane domain then it will be excreted through the inner membrane. If a protein with a signal peptide has a transmembrane domain then it will become inserted into the membrane.[200] All signal peptides have a three region structure, the amino (N), the hydrophobic (H) and the carboxy (C) with a weak consensus pattern specifying the cleavage site.[201] Signal peptides are divided into classes on the basis of variation of structure of the N, H and C regions, structure of the cleavage site and different propensities for amino acids.[202]

Many approaches have been taken to try and predict subcellular location from signal peptides and cleavage variations. The different amino acid propensities of N, H and C regions for different classes can be identified by multivariate analysis of the individual amino acids.[203] A wide range of characteristics of amino acid properties has been determined and the similarities/dissimilarities in the property profiles for different signal peptide classes were compared. Initially this method was applied to just *E. coli* with some success but later expansion to Gram positive bacteria was less successful and varied greatly from species to species.[204] There were some factors though such as charge, length, sidechain hydrophobicity and volume that proved reasonably reliable factors that could be used as part of possible new techniques. The prediction of cleavage sites and inference of subcellular location has proved more fruitful than amino acid composition based methods, with prediction as high as 96%.[205,206]

6 Applications to the Discovery of Vaccines

Ultimately, the utilitarian value of the many techniques described above will need to be demonstrated through their usefulness in experimental vaccine discovery programmes. All of the methods we have adduced focus primarily on the discovery of T cell epitopes, which can prove useful, amongst other things, as diagnostic markers of microbial infection and as the potential basis of epitope vaccines. Many workers have, in recent years, used computational methods as part of their strategy for the identification of both Class I and Class II restricted T cell epitopes, but it is outside the scope of the present work to review these studies in detail. However, it is certainly encouraging that many experimental immunologists are now beginning to see the need for informatics techniques. One of the key problems they face is the information explosion poised to sweep over immunology. Computer-based data and knowledge management, as manifest in the development of novel databases[27,104,119] and predictive methods,[97,125,163] is essential if this data deluge is not to overwhelm the immunologists in the post-genomic era. What the field currently lacks are any very convincing comparative studies of the performance of these different algorithms. Two papers that come close to performing such an analysis reach very different conclusions. Lu and Celis[207] used two publically available prediction algorithms – BIMAS[22] and SYFPEITHI[27] – to identify B7 restricted CTL epitopes within the carcinoembryonic antigen (CEA), yielding three candidate peptides that were tested for T cell responses. One CEA peptide: IPQQHTQVL, efficiently induced a CTL response. They concluded that 'our strategy of identifying MHC Class

I-restricted CTL epitopes without the need of peptide/HLA-binding assays provides a convenient and cost-saving alternative approach to previous methods'. In contrast to this highly upbeat message, Andersen *et al.*[208] analysed the experimental binding of 84 peptides selected using the presence of allele-dependent peptide binding motifs. Observed binding was compared with results obtained from the same two algorithms used by Lu and Celis. The authors concluded that no strong correlation exists between actual and predicted binding using these algorithms. Moreover, they also found a high number of false negatives when using the BIMAS or SYFPEITHI algorithms compared to simple scanning for primary anchor residues. Andersen *et al.* concluded that 'the peptide binding assay remains an important step in the identification of CTL epitopes which can not be substituted by predictive algorithms'. Thus it is clear that there is a need to produce more accurate prediction algorithms, which cover more Class I and Class II alleles in more species. Yet, for these improved methodologies to be ultimately effective – that is to say that they are taken up and used routinely by experimental immunologists – these methods must also be tested rigorously for a sufficiently large number of peptides that their accuracy can be shown to work to statistical significance.

7 Conclusion

As we have seen, the accurate prediction of epitopes is one weapon that can be used to combat the impending flood of new post-genomic information, but there are many more, some of which we have mentioned in passing, others to which we have made no allusion. These include methods for the prediction of numerous properties of the humoural immune system such as B cell epitopes or neutralising Abs binding sites, the prediction of whole protein, glycoprotein, or lipoprotein antigens, or clever strategies for presentation of defined epitopes in protein vectors. Indeed, the whole area of immunogenicity includes not only prediction of T cell and B cell epitopes, but also many other things beside. We have not mentioned the role that the rational optimization of epitopes, in terms of MHC binding or TCR recognition, can play in improving immunogenic or protective qualities of epitope-based vaccines, or how the use of computer based screening can help in the search for promiscuous peptides, able to bind many different allleles, which is a very important concern, particularly for Class II or the part that variability of antigen sequence can play. There is much that informatic techniques can offer, including solutions, or partial solutions, to those outstanding problems mentioned above.

Another important observation to arise from many of the studies mentioned here, but primarily from application of QSAR and MD methods, is the emphasis placed on the important role of non-anchor residues in influencing the energetics of peptide–MHC binding. It is clear that anchor residues alone cannot account for peptide binding. Rather it is the combination of all amino acids within the peptide that ultimately determines the observed affinity of binding.

It is paradoxical that despite the brilliant insights of immunologists over the decades, there remain many fundamental problems of the immune system that are poorly understood. Computational methods can aid this search for understanding. Indeed, it may be essential to overcome the information overload that is about to break over the subject. Because of the diversity of the immune system, such computational approaches will be neccessary in the discovery and design of both individualized and population-based vaccines. Computational simulation has already proved useful, for example in T cell epitope mapping, to support more efficient experimentation and discovery. While it is clear that bacterial pathogens have evolved many ingenious methods to invade the human host successfully, it is also clear that many seemingly diverse pathogens can share common virulence traits. This will be of great use when designing novel compounds to combat diseases. As each new genome sequenced increases our knowledge of microbial pathogenesis, so the number of targets available for therapeutic research should likewise increase.

The next step will come from closer connections between immunoinformaticians and experimentalists seeking to discover new vaccines, both academic and commercial, conducted under a collaborative or consultant regime. In such a situation, work would progress through a cyclical process of using and refining models and experiments, at each stage moving closer towards a common goal of effective, cost-efficient vaccine development. This is certainly the focus and objective of this group. The allied subjects of bioinformatics and molecular modelling have proved their worth time and again in the search for new drugs and drug targets. The time is approaching when they will do the same for vaccine design. Methods that allow us to predict accurately epitopes or immunogenic proteins, or to eliminate virulence factor genes from individual bacteria, will prove to be crucial tools for the vaccinologist of tomorrow.

Acknowledgements

We should like to thank the following for their help and stimulating discussions: Prof Peter Beverley, Dr Persephone Borrow, Dr Kevin Rigley, Dr Simon Wong, Dr Helen Bodmer, Dr Sam Hou, Dr Elma Tchillian, Mr Josef Walker, and Dr Vladimir Brusic. We should also like to thank all our colleagues and co-workers at the Edward Jenner Institute for Vaccine Research and Institute of Animal Health, Compton for their close and supportive collaboration.

References

1. E.G. Krug, G.K. Sharma and R. Lozano, Accidental death, *Am. J. Public Health*, 2000, **90**, 523–526.
2. A.E. Van der Bogaard and E.E. Stobberingh, Epidemiology of resistance to antibiotics. Links between animals and humans, *Int. J. Antimicrob. Agents*, 2000, **14**, 327–335.
3. The World Health Organisation Report on Infectious Diseases 2000. http://www.who.int/infectious-disease-report/2000/index.html.

4. H. Ochman, J.G. Lawrence and E.A. Groisman, Lateral gene transfer and the nature of bacterial innovation, *Nature*, 2000, **405**, 299–304.
5. A.K. Nussbaum, T.P. Dick, W. Keilholz, M. Schirle, S. Stevanovic, K. Dietz, W. Heinemeyer, M. Groll, D.H. Wolf, R. Huber, H.G. Rammensee and H. Schild, Cleavage motifs of the yeast 20S proteasome beta subunits deduced from digests of enolase 1, *Proc. Nat. Acad. Sci. USA*, 1998, **95**, 12504–12509.
6. H.G. Holzhutter, C. Frommel and P.M. Kloetzel, A theoretical approach towards the identification of cleavage-determining amino acid motifs of the 20S proteasome, *J. Mol. Biol.*, 1999, **286**,1251–1265.
7. Y. Altuvia and H. Margalit, Sequence signals for generation of antigenic peptides by the proteasome: implications for proteasomal cleavage mechanism, *J. Mol. Biol.*, 2000, **295**, 879–890.
8. C. Kuttler, A.K. Nussbaum, T.P. Dick, H.G. Rammensee, H. Schild and K.P. Hadeler, An algorithm for the prediction of proteasomal cleavages, *J. Mol. Biol.*, 2000, **298**, 417–429.
9. A.K. Nussbaum, C. Kuttler, K.P. Hadeler, H.G. Rammensee and H. Schild, PAProC: a prediction algorithm for proteasomal cleavages available on the WWW, *Immunogenetics*, 2001, **53**, 87–94.
10. S. Daniel, V. Brusic, S. Caillat-Zucman, N. Petrovsky, L. Harrison, D. Riganelli, F. Sinigaglia, F. Gallazzi, J.,Hammer and P.M. van Endert, Relationship between peptide selectivities of human transporters associated with antigen processing and HLA class I molecules, *Immunology*, 1998, **161**, 617–624.
11. V. Brusic, P. van Endert, J. Zeleznikow, S. Daniel, J. Hammer and N. Petrovsky, A neural network model approach to the study of human TAP transporter, *In Silico Biol.*, 1999, **1**, 109–121.
12. J. Lu, P.J. Wettstein, Y. Higashimoto, E. Appella,and E. Celis, TAP-independent presentation of ctl epitopes by trojan antigens, *J. Immunol.*, 2001, **166**, 7063–7071.
13. B.C. Gil-Torregrosa, A.R. Castano, D. Lopez and M. Del Val, Generation of MHC class I peptide antigens by protein processing in the secretory route by furin, *Traffic*, 2000, **1**, 641–651.
14. H.A. Chapman, Endosomal proteolysis and MHC class II function, *Curr. Opin. Immunol.*, 1998, **10**, 93–102.
15. J. Ruppert, J. Sidney, E. Celis, R.T. Kubo, H.M. Grey and A. Sette, Prominent role of secondary anchor residues in peptide binding to HLA-A2.1 molecules, *Cell*, 1994, **74**, 929–934.
16. A. Sette, J. Sidney, M.F. del Guercio, S. Southwood, J. Ruppert, C. Dahlberg, H.M. Grey and R.T. Kubo, Peptide binding to the most frequent HLA-A class I alleles measured by quantitative molecular binding assays, *Mol. Immunol.*, 1994, **31**, 813–820.
17. J. Sidney, C. Oseroff, M.F. del Guercio, S. Southwood, J.I. Krieger, G.Y. Ishioka, K. Sakaguchi, E. Appella and A. Sette, Definition of a DQ3.1-specific binding motif, *J. Immunol.*, 1994, **152**, 4516–4523.
18. K.W. Marshall, A.F. Liu, J. Canales, B. Perahia, B. Jorgensen, R.D. Gantzos, B. Aguilar, B. Devaux and J.B. Rothbard, Role of the polymorphic residues in HLA-DR molecules in allele-specific binding of peptide ligands. *J. Immunol.*, 1994, **152**, 4946–4953.
19. S.H. van der Burg, M.J. Visseren, R.M. Brandt, W.M. Kast and C.J. Melief, Immunogenicity of peptides bound to MHC class I molecules depends on the MHC-peptide complex stability, *J. Immunol.*, 1996, **156**, 3308–3315.
20. G.J. ten Bosch, J.H. Kessler, A.M. Joosten, A.A. Bres-Vloemans, A. Geluk, B.C.

Godthelp, J. van Bergen, C.J. Melief and O.C. Leeksma, A BCR-ABL oncoprotein p210b2a2 fusion region sequence is recognized by HLA-DR2a restricted cytotoxic T lymphocytes and presented by HLA-DR matched cells transfected with an Ii(b2a2) construct, *Blood*, 1999, **94**, 10381–10387.

21. K.C. Parker, M. DiBrino, L. Hull and J.E. Coligan, The beta 2-microglobulin dissociation rate is an accurate measure of the stability of MHC class I heterotrimers and depends on which peptide is bound, *J. Immunol.*, 1992, **149**, 1896–1903.

22. K.C. Parker, M.A. Bednarek and J.E. Coligan, Scheme for ranking potential HLA-A2 binding peptides based on independent binding of individual peptide side-chains, *J. Immunol.*, 1994, **152**, 163–170.

23. M. DiBrino, K.C. Parker, J. Shiloach, R.V. Turner, T. Tsuchida, M. Garfield, W.E. Biddison and J.E. Coligan, Endogenous peptides with distinct amino acid anchor residue motifs bind to HLA-A1 and HLA-B8, *J. Immunol.*, 1994, **152**, 620–629.

24. V. Brusic, G. Rudy, A.P. Kyne and L.C. Harrison, MHCPEP – a database of MHC-binding peptides: update 1995, *Nucleic Acids Res.*, 1996, **24**, 242–244.

25. A. Sette, A. Vitiello, B. Reherman, P. Fowler, R. Nayersina, W.M. Kast, C.J. Melief, C. Oseroff, L. Yuan and J. Ruppert, The relationship between class I binding affinity and immunogenicity of potential cytotoxic T cell epitopes, *J. Immunol.*, 1994, **153**, 5586–5592.

26. A. Sette, S. Buus, E. Appella, J.A. Smith, R. Chesnut, C. Miles, S.M. Colon and H.M. Grey, Prediction of major histocompatibility complex binding regions of protein antigens by sequence pattern analysis, *Proc. Nat. Acad. Sci. USA*, 1989, **86**, 3296–3300.

27. H. Rammensee, J. Bachmann, N.P. Emmerich, O.A. Bachor and S. Stevanovic, SYFPEITHI: database for MHC ligands and peptide motifs, *Immunogenetics*, 1999, **50**, 213–219.

28. Y. Becker, Computer simulations of proteolysis of Marburg and Ebola-Zaire filovirus coded proteins to generate nonapeptides with motifs of known HLA class I haplotypes and detection of antigenic domains in the viral glycoproteins, *Virus Genes*, 1996, **13**, 189–201.

29. Y. Becker, Computer simulations to identify in polyproteins of FMDV OK1 and A12 strains putative nonapeptides with amino acid motifs for binding to BoLA class I A11 and A20 haplotype molecules, *Virus Genes*, 1997, **14**, 123–129.

30. J. D'Amaro, J.G. Houbiers, J.W. Drijfhout, R.M. Brandt, R. Schipper, J.N. Bavinck, C.J. Melief and Kast, A computer program for predicting possible cytotoxic T lymphocyte epitopes based on HLA class I peptide-binding motifs, *Hum. Immunol.*, 1995, **43**, 13–18.

31. F. Sinigaglia and J. Hammer, Motifs and supermotifs for MHC class II binding peptides, *Sette A. Exp. Med.*, 1995, **181**, 449–451.

32. C. Zhang, J.L. Cornette, J.A. Berzofsky and C. DeLisi, The organization of human leucocyte antigen class I epitopes in HIV genome products: implications for HIV evolution and vaccine design, *Vaccine*, 1997, **15**, 1291–1302.

33. P.A. Reay, R.M. Kantor and M.M. Davis, Use of global amino acid replacements to define the requirements for MHC binding and T cell recognition of moth cytochrome c (93–103), *J. Immunol.*, 1994, **152**, 3946–3957.

34. J.B. Rothbard, K. Marshall, K.J. Wilson, L. Fugger and D. Zaller, Prediction of peptide affinity to HLA DRB1*0401, *Int. Arch. Allergy. Immunol.*, 1994, **105**, 1–7.

35. K.W.Marshall, K.J. Wilson, J. Liang, A. Woods, D. Zaller and J.B. Rothbard, Prediction of peptide affinity to HLA DRB1*0401, *J. Immunol.*, 1995, **154**, 5927–5933.

36. K.W. Marshall, K.J. Wilson, J. Liang, A. Woods, D. Zaller and J.B. Rothbard,

Prediction of peptide affinity to HLA DR molecules, *Biomed. Pept . Proteins Nucleic Acids*, 1995, **1**, 157–162.

37. A. Stryhn, L.O. Pedersen, T. Romme, C.B. Holm, A. Holm and S. Buus, Peptide binding specificity of major histocompatibility complex class I resolved into an array of apparently independent subspecificities: quantitation by peptide libraries and improved prediction of binding, *Eur. J. Immunol.*, 1996, **26**, 1911–1818.

38. J. Stevens, K.H. Wiesmuller, P. Walden and E. Joly, Peptide length preferences for rat and mouse MHC class I molecules using random peptide libraries, *Eur. J. Immunol.*, 1998, **28**, 1272–1279.

39. J. Stevens, K.H. Wiesmuller, P.J. Barker, P. Walden, G.W. Butcher and E. Joly, Efficient generation of major histocompatibility complex class I-peptide complexes using synthetic peptide libraries, *J. Biol. Chem.*, 1998, **273**, 2874–2884.

40. Y. Zhao, B. Gran, C. Pinilla, S. Markovic-Plese, B. Hemmer, A. Tzou, L.W. Whitney, W.E. Biddison, R. Martin and R. Simon, Combinatorial peptide libraries and biometric score matrices permit the quantitative analysis of specific and degenerate interactions between clonotypic TCR and MHC peptide ligands, *J. Immunol.*, 2001, **167**, 2130–2141.

41. C. Pinilla, V. Rubio-Godoy, V. Dutoit, P. Guillaume, R. Simon, Y. Zhao, R.A. Houghten, J.C. Cerottini, P. Romero and D. Valmori, Combinatorial peptide libraries as an alternative approach to the identification of ligands for tumor-reactive cytolytic T lymphocytes, *Cancer Res.*, 2001, **61**, 5153–5160.

42. K. Udaka, K.H. Wiesmuller, S. Kienle, G. Jung, H. Tamamura, H. Yamagishi, K. Okumura, P. Walden, T. Suto and T. Kawasaki, An automated prediction of MHC class I-binding peptides based on positional scanning with peptide libraries, *Immunogenetics*, 2000, **51**, 816–828.

43. K.C. Parker, M. Shields, M. DiBrino, A. Brooks and J.E. Coligan, Peptide binding to MHC class I molecules: implications for antigenic peptide prediction, *Immunol. Res.*, 1995, **14**, 34–57.

44. K.C. Parker, M.A. Bednarek, L.K. Hull, U. Utz, B. Cunningham, H.J. Zweerink, W.E. Biddison and J.E. Coligan, Sequence motifs important for peptide binding to the human MHC class I molecule, HLA-A2, *J. Immunol.*, 1992, **149**, 3580–3587.

45. K.C. Parker, B.M. Carreno, L. Sestak, U. Utz, W.E. Biddison and J.E. Coligan, Peptide binding to HLA-A2 and HLA-B27 isolated from *Escherichia coli*. Reconstitution of HLA-A2 and HLA-B27 heavy chain/beta 2-microglobulin complexes requires specific peptides, *J. Biol. Chem.*, 1992, **267**, 5451–5459.

46. M. DiBrino, K.C. Parker, J. Shiloach, M. Knierman, J. Lukszo, R.V. Turner, W.E. Biddison and J.E. Coligan, Endogenous peptides bound to HLA-A3 possess a specific combination of anchor residues that permit identification of potential antigenic peptides, *Proc. Nat. Acad. Sci. USA*, 1993, **90**, 1508–1512.

47. M. DiBrino, K.C. Parker, D.H. Margulies, J. Shiloach, R.V. Turner, W.E. Biddison and J.E. Coligan, The HLA-B14 peptide binding site can accommodate peptides with different combinations of anchor residues, *J. Biol. Chem.*, 1994, **269**, 32426–32434.

48. K.C. Parker, W.E. Biddison and J.E. Coligan, Pocket mutations of HLA-B27 show that anchor residues act cumulatively to stabilize peptide binding, *Biochemistry*, 1994, **33**, 7736–7743.

49. M. DiBrino, K.C. Parker, D.H. Margulies, J. Shiloach, R.V. Turner, W.E. Biddison and J.E. Coligan, Identification of the peptide binding motif for HLA-B44, one of the most common HLA-B alleles in the Caucasian population, *Biochemistry*, 1995, **34**, 10130–10138.

50. M. DiBrino, T. Tsuchida, R.V. Turner, K.C. Parker, J.E. Coligan and W.E. Biddison,

HLA-A1 and HLA-A3 T cell epitopes derived from influenza virus proteins predicted from peptide binding motifs, *J. Immunol.*, 1993, **151**, 5930–5935.

51. K. Honma, K.C. Parker, K.G. Becker, H.F. McFarland, J.E. Coligan and W.E. Biddison, Identification of an epitope derived from human proteolipid protein that can induce autoreactive CD8+ cytotoxic T lymphocytes restricted by HLA-A3: evidence for cross-reactivity with an environmental microorganism, *J. Neuroimmunol.*, 1997, **73**, 7–14.

52. M.A. Alexander-Miller, K.C. Parker, T. Tsukui, C.D. Pendleton, J.E. Coligan and J.A. Berzofsky, Molecular analysis of presentation by HLA-A2.1 of a promiscuously binding V3 loop peptide from the HIV-envelope protein to human cytotoxic T lymphocytes, *Int. Immunol.*, 1996, **8**, 641–649.

53. C.R. Alves, L.C. Pontes de Carvalho, A.L. Souza and S.G.De Simone, A strategy for the identification of T-cell epitopes on Leishmania cysteine proteinases, *Cytobios*, 2001, **104**, 33–41.

54. F. Borras-Cuesta, J. Golvano, M. Garcia-Granero, P. Sarobe, J. Riezu-Boj, E. Huarte and J. Lasarte, Specific and general HLA-DR binding motifs: comparison of algorithms, *Hum. Immunol.*, 2000, **61**, 266–278.

55. A.S. De Groot, M. Clerici, A. Hosmalin, S.H. Hughes, D. Barnd, C.W. Hendrix, R. Houghten, G.M. Shearer and J.A. Berzofsky, Human immunodeficiency virus reverse transcriptase T helper epitopes identified in mice and humans: correlation with a cytotoxic T cell epitope, *J. Infect. Dis.*, 1991, **164**, 1058–1065.

56. G.E. Meister, C.G. Roberts, J.A. Berzofsky and A.S. De Groot, Two novel T cell epitope prediction algorithms based on MHC-binding motifs; comparison of predicted and published epitopes from *Mycobacterium tuberculosis* and HIV protein sequences, *Vaccine*, 1995, **13**, 581–591.

57. C.G. Roberts, G.E. Meister, B.M. Jesdale, J. Lieberman, J.A. Berzofsky and A.S. De Groot, Prediction of HIV peptide epitopes by a novel algorithm, *AIDS Res Hum. Retroviruses*, 1996, **12**, 593–610.

58. A.S. De Groot, B.M. Jesdale, E. Szu, J.R. Schafer, R.M. Chicz and G. Deocampo, An interactive Web site providing major histocompatibility ligand predictions: application to HIV research, *AIDS Res. Hum. Retroviruses*, 1997, **13**, 529–531.

59. J.R. Schafer, B.M. Jesdale, J.A. George, N.M. Kouttab and A.S. De Groot, Prediction of well-conserved HIV–1 ligands using a matrix-based algorithm, EpiMatrix, *Vaccine*, 1998, **16**, 1880–1884.

60. A.S. De Groot and F.G. Rothman, In silico predictions; in vivo veritas, *Nat. Biotechnol.*, 1999, **17**, 533–534.

61. X. Jin, C.G. Roberts, D.F. Nixon, J.T. Safrit, L.Q. Zhang, Y.X. Huang, N. Bhardwaj, B.M. Jesdale, A.S. De Groot and R.A. Koup, Identification of subdominant cytotoxic T lymphocyte epitopes encoded by autologous HIV type 1 sequences, using dendritic cell stimulation and computer-driven algorithm, *AIDS Res. Hum. Retroviruses*, 2000, **16**, 67–76.

62. A.S. De Groot, C. Saint-Aubin, A. Bosma, H. Sbai, J. Rayner and W. Martin, Rapid determination of hla b*07 ligands from the west nile virus ny99 genome, *Emerg. Infect. Dis.*, 2001, **7**, 706–713.

63. A.S. De Groot, A. Bosma, N. Chinai, J. Frost, B.M. Jesdale, M.A. Gonzalez, W. Martin and C. Saint-Aubin, From genome to vaccine: in silico predictions, ex vivo verification, *Vaccine*, 2001, **19**, 4385–4395.

64. K.B. Bond, B. Sriwanthana, T.W. Hodge, A.S. De Groot, T.D. Mastro, N.L. Young, N. Promadej, J.D. Altman, K. Limpakarnjanarat and J.M. McNicholl, An HLA-directed molecular and bioinformatics approach identifies new HLA-A11 HIV-1

subtype E cytotoxic T lymphocyte epitopes in HIV-1-infected Thais. *AIDS Res. Hum. Retroviruses*, 2001, **17**, 703–717.

65. J. Hammer, B. Takacs and F. Sinigaglia, Identification of a motif for HLA-DR1 binding peptides using M13 display libraries, *J. Exp. Med.*, 1992, **176**, 1007–1013.
66. J. Hammer, P. Valsasnini, K. Tolba, D. Bolin, J. Higelin, B. Takacs and F. Sinigaglia, Promiscuous and allele-specific anchors in HLA-DR-binding peptides, *Cell*, 1993, **74**, 197–203.
67. J. Hammer, C. Belunis, D. Bolin, J. Papadopoulos, R. Walsky, J. Higelin, W. Danho, F. Sinigaglia and Z.A. Nagy, High-affinity binding of short peptides to major histocompatibility complex class II molecules by anchor combinations, *Proc. Nat. Acad. Sci. USA*, 1994, **91**, 4456–4460.
68. F. Sinigaglia and J. Hammer, Rules for peptide binding to MHC class II molecules, *APMIS*, 1994, **102**, 241–248.
69. F. Sinigaglia and J. Hammer, Defining rules for the peptide-MHC class II interaction, *Curr. Opin. Immunol.*, 1994, **6**, 52–56.
70. J. Hammer, E. Bono, F. Gallazzi, C. Belunis, Z. Nagy and F. Sinigaglia, Precise prediction of major histocompatibility complex class II-peptide interaction based on peptide side chain scanning, *J. Exp. Med.*, 1994, **180**, 2353–2358.
71. F. Sinigaglia and J. Hammer, Predicting major histocompatibility complex-binding sequences within protein antigens, *Biochem. Soc. Trans.*, 1995, **23**, 675–677.
72. J. Hammer, F. Gallazzi, E. Bono, R.W. Karr, J. Guenot, P. Valsasnini, Z.A. Nagy and F. Sinigaglia, Peptide binding specificity of HLA-DR4 molecules: correlation with rheumatoid arthritis association, *J. Exp. Med.*, 1995, **181**, 1847–1855.
73. J. Hammer, New methods to predict MHC-binding sequences within protein antigens, *Curr. Opin. Immunol.*, 1995, **7**, 263–269.
74. F. Sinigaglia and J. Hammer, Motifs and supermotifs for MHC class II binding peptides, *J. Exp. Med.*, 1995, **181**, 449–451.
75. J.M. Calvo-Calle, J. Hammer, F. Sinigaglia, P. Clavijo, Z.R. Moya-Castro and E.H. Nardin, Binding of malaria T cell epitopes to DR and DQ molecules in vitro correlates with immunogenicity in vivo: identification of a universal T cell epitope in the Plasmodium falciparum circumsporozoite protein, *J. Immunol.*, 1997, **159**, 1362–1373.
76. J. Hammer, T. Sturniolo and F. Sinigaglia, HLA class II peptide binding specificity and autoimmunity, *Adv. Immunol.*, 1997, **66**, 67–100.
77. L. Raddrizzani, T. Sturniolo, J. Guenot, E. Bono, F. Gallazzi, Z.A. Nagy, F. Sinigaglia and J. Hammer, Different modes of peptide interaction enable HLA-DQ and HLA-DR molecules to bind diverse peptide repertoires, *J. Immunol.*, 1997, **159**, 703–711.
78. G.J. McColl, J. Hammer and L.C. Harrison, Absence of peripheral blood T cell responses to 'shared epitope' containing peptides in recent onset rheumatoid arthritis, *Ann. Rheum. Dis.*, 1997, **56**, 240–246.
79. L.C. Harrison, M.C. Honeyman, S. Trembleau, S. Gregori, F. Gallazzi, P. Augstein, V. Brusic, J. Hammer and L. Adorini, A peptide-binding motif for I-A(g7), the class II major histocompatibility complex (MHC) molecule of NOD and Biozzi AB/H mice, *J. Exp. Med.*, 1997, **185**, 10113–10121.
80. S. Gregori, S. Trembleau, G. Penna, F. Gallazzi, J. Hammer, G.K. Papadopoulos and L. Adorini, A peptide binding motif for I-Eg7, the MHC class II molecule that protects E alpha-transgenic nonobese diabetic mice from autoimmune diabetes, *J. Immunol.*, 1999, **162**, 6630–6640.
81. S. Gregori, E. Bono, F. Gallazzi, J. Hammer, L.C. Harrison and L. Adorini, The motif

for peptide binding to the insulin-dependent diabetes mellitus-associated class II MHC molecule I-Ag7 validated by phage display library, *Int. Immunol.*, 2000, **12**, 493–503.

82. W.W. Kwok, J.A., Gebe, A. Liu, S. Agar, N. Ptacek, J. Hammer, D.M. Koelle and G.T. Nepom, Rapid epitope identification from complex class-II-restricted T-cell antigens, *Trends Immunol.*, 2001, **22**, 583–588.

83. L. Raddrizzani and J. Hammer, Epitope scanning using virtual matrix-based algorithms, *Brief. Bioinform.*, 2000, **1**, 179–189.

84. B. Cochlovius, M. Stassar, O. Christ, L. Raddrizzani, J. Hammer, I. Mytilineos and M. Zoller, In vitro and in vivo induction of a Th cell response toward peptides of the melanoma-associated glycoprotein 100 protein selected by the TEPITOPE program, *J. Immunol.*, 2000, **165**, 4731–4741.

85. M.J. Stassar, L. Raddrizzani, J. Hammer and M. Zoller, T-helper cell-response to MHC class II-binding peptides of the renal cell carcinoma-associated antigen RAGE-1, *Immunobiology*, 2001, **203**, 743–755.

86. C. de Lalla, T. Sturniolo, L. Abbruzzese, J. Hammer, A. Sidoli, F. Sinigaglia and P. Panina-Bordignon, Cutting edge: identification of novel T cell epitopes in Lol p5a by a computational prediction, *J. Immunol.*, 1999, **163**, 1725–1729.

87. L. Raddrizzani, E. Bono, A.B. Vogt, H. Kropshofer, F. Gallazzi, T. Sturniolo, G.J. Hammerling, F. Sinigaglia and J. Hammer, Identification of destabilizing residues in HLA class II-selected bacteriophage display libraries edited by HLA-DM, *Eur. J. Immunol.*, 1999, **29**, 660–668.

88. T. Sturniolo, E. Bono, J. Ding, L. Raddrizzani, O. Tuereci, U. Sahin, M. Braxenthaler, F. Gallazzi, M.P. Protti, F. Sinigaglia and J. Hammer, Generation of tissue-specific and promiscuous HLA ligand databases using DNA microarrays and virtual HLA class II matrices, *Nat. Biotechnol.*, 1999, **17**, 555–561.

89. L.R. Bisset and W. Fierz, Using a neural network to identify potential HLA-DR1 binding sites within proteins, *J. Mol. Recognit.*, 1993, **6**, 41–48.

90. H.P. Adams and J.A. Koziol, Prediction of binding to MHC class I molecules, *J. Immunol. Methods*, 1995, **185**, 181–190.

91. K. Gulukota, J. Sidney, A. Sette and C. DeLisi, Two complementary methods for predicting peptides binding major histocompatibility complex molecules, *J. Mol. Biol.*, 1997, **267**, 1258–1267.

92. K. Gulukota, C. DeLisi, Neural network method for predicting peptides that bind major histocompatibility complex molecules, *Methods Mol. Biol.*, 2001, **156**, 201–209.

93. V. Brusic, C. Schonbach, M. Takiguchi, V. Ciesielski and L.C. Harrison, Application of genetic search in derivation of matrix models of peptide binding to MHC molecules, *Proc. Int. Conf. Intell. Syst. Mol. Biol.*, 1997, **5**, 75–83.

94. L.C. Harrison, M.C. Honeyman, S. Trembleau, S. Gregori, F. Gallazzi, P. Augstein, V. Brusic, J. Hammer and L. Adorini, A peptide-binding motif for I-A(g7), the class II major histocompatibility complex (MHC) molecule of NOD and Biozzi AB/H mice, *J. Exp. Med.*, 1997, **185**, 1013–1021.

95. M.C. Honeyman, V. Brusic and L.C. Harrison, Strategies for identifying and predicting islet autoantigen T-cell epitopes in insulin-dependent diabetes mellitus, *Ann. Med.*, 1997, **29**, 401–404.

96. M.C. Honeyman, V. Brusic, N.L. Stone and L.C. Harrison, Neural network-based prediction of candidate T-cell epitopes, *Nat. Biotechnol.*, 1998, **16**, 966–969.

97. V. Brusic, G. Rudy, G. Honeyman, J. Hammer and L. Harrison, Prediction of MHC class II-binding peptides using an evolutionary algorithm and artificial neural network, *Bioinformatics*, 1998, **14**, 121–130.

98. H.M. Zarour, W.J. Storkus, V. Brusic, E. Williams and J.M. Kirkwood, NY-ESO-1 encodes DRB1*0401-restricted epitopes recognized by melanoma-reactive CD4+ T cells, *Cancer Res.*, 2000, **60**, 4946–4952.

99. H.M. Zarour, J.M. Kirkwood, L.S. Kierstead, W. Herr, V. Brusic, C.L. Slingluff Jr, J. Sidney, A. Sette and W.J. Storkus, Melan-A/MART-1(51–73) represents an immunogenic HLA-DR4-restricted epitope recognized by melanoma-reactive CD4(+) T cells, *Proc. Nat. Acad. Sci. USA*, 2000, **97**, 400–405.

100. V. Brusic, K. Bucci, C. Schonbach, N. Petrovsky, J. Zeleznikow and J.W. Kazura, Efficient discovery of immune response targets by cyclical refinement of QSAR models of peptide binding, *J. Mol. Graph. Model.*, 2001, **19**, 405–411.

101. V. Brusic, G. Rudy and L.C. Harrison MHCPEP, a database of MHC-binding peptides: update 1997, *Nucleic Acids Res.*, 1998, **26**, 368–371.

102. V. Brusic, G. Rudy, A.P. Kyne and L.C. Harrison MHCPEP, a database of MHC-binding peptides: update 1996, *Nucleic Acids Res.*, 1997, **25**, 269–271.

103. V. Brusic, G. Rudy and L.C. Harrison, MHCPEP: a database of MHC-binding peptides, *Nucleic Acids Res.*, 1994, **22**, 3663–3665.

104. C. Schonbach, J.L. Koh, X. Sheng, L. Wong and V. Brusic, FIMM, a database of functional molecular immunology, *Nucleic Acids Res.*, 2000, **28**, 222–224.

105. V. Brusic, J. Zeleznikow and N. Petrovsky, Molecular immunology databases and data repositories, *J. Immunol. Methods*, 2000, **238**, 17–28.

106. M. Milik, D. Sauer, A.P. Brunmark, L. Yuan, A. Vitiello, M.R. Jackson, P.A. Peterson, J. Skolnick and C.A. Glass, Application of an artificial neural network to predict specific class I MHC binding peptide sequences, *Nat. Biotechnol.*, 1998, **16**, 753–756.

107. C.J. Savoie, N. Kamikawaji, T. Sasazuki and S. Kuhara, Use of BONSAI decision trees for the identification of potential MHC class I peptide epitope motifs, *Pac. Symp. Biocomput.*, 1999, 182–189.

108. H. Mamitsuka, Predicting peptides that bind to MHC molecules using supervised learning of hidden Markov models, *Proteins*, 1998, **33**, 460–474.

109. R.R. Mallios, Multiple regression analysis suggests motifs for class II MHC binding, *J. Theor. Biol.*, 1994, **166**, 167–172.

110. R.R. Mallios, An iterative algorithm for converting a class II MHC binding motif into a quantitative predictive model, *Comput. Appl. Biosci.*, 1997, **13**, 211–215.

111. R.R. Mallios, Iterative stepwise discriminant analysis: a meta-algorithm for detecting quantitative sequence motifs, *J. Comput. Biol.*, 1998, **5**, 703–711.

112. R.R. Mallios, Class II MHC quantitative binding motifs derived from a large molecular database with a versatile iterative stepwise discriminant analysis meta-algorithm, *Bioinformatics*, 1999, **15**, 432–439.

113. R.R. Mallios, Predicting class II MHC/peptide multi-level binding with an iterative stepwise discriminant analysis meta-algorithm, *Bioinformatics*, 2001, **17**, 942–948.

114. R.R. Mallios, Predicting the probability of helper T cell immunodominant sites through discriminant analysis, *Ann. Clin. Biochem.*, 1993, **30**, 152–156.

115. A. Chersi, F. di Modugno and L. Rosano, Flexibility of amino acid residues at position four of nonapeptides enhances their binding to human leucocyte antigen (HLA) molecules, *Z. Naturforsch.*, 2000, **55**, 109–114.

116. C. Bologa, D. Drugarin and Z. Simon, Quantitative structure–activity relations by the MTD-method for binding of nonapeptides to the HLA-A2.1 molecule, *Roum. Arch. Microbiol. Immunol.*, 1995, **54**, 3–14.

117. P. Rovero, D. Riganelli, D. Fruci, S. Vigano, S. Pegoraro, R. Revoltella, G. Greco, R. Butler, S. Clementi and N. Tanigaki, The importance of secondary anchor residue

motifs of HLA class I proteins: a chemometric approach, *Mol. Immunol.*, 1994, **31**, 549–554.
118. S.L. Lauemoller, A. Holm, J. Hilden, S. Brunak, M. Holst Nissen, A. Stryhn, L. Ostergaard Pedersen and S. Buus, Quantitative predictions of peptide binding to MHC class I molecules using specificity matrices and anchor-stratified calibrations, *Tissue Antigens*, 2001, **57**, 405–414.
119. M.J. Blythe, I.A. Doytchinova and D.R. Flower, Jenpep: A database of quantitative functional peptide data for Immunology, *Bioinformatics*, 2002, **18**, 434–439.
120. R.D. Crammer, III, D.E. Patterson and J.D. Bunce, Comparative Molecular Field Analysis (CoMFA). 1. Effect of shape on binding of steroids to carrier proteins, *J. Am. Chem. Soc.*, 1988, **110**, 5959–5967.
121. G. Klebe, U. Abraham and T. Mietzner, Molecular Similarity Indices in a Comparative Analysis (CoMSIA) of drug molecules to correlate and predict their biological activity, *J. Med. Chem.*, 1994, **37**, 4130–4146.
122. G. Klebe and U. Abraham, Comparative Molecular Similarity Index Analysis (CoMSIA) to study hydrogen-bonding properties and to score combinatorial libraries, *J. Comput.-Aided Mol. Design*, 1999, **13**, 1–10.
123. M. Böhm, J. Stürzebecher and G. Klebe, Three-dimensional quantitative structure–activity relationship analyses using comparative molecular field analysis and comparative molecular similarity indices analysis to elucidate selectivity differences of inhibitors binding to trypsin, thrombin, and factor Xa, *J. Med. Chem.*, 1999, **42**, 458–477.
124. G. Folkers, A. Merz and D. Rognan, *CoMFA: Scope and Limitations in 3D QSAR in Drug Design*, ed. H. Kubinyi, ESCOM, Leiden, 1993, pp. 583–618.
125. I. Doytchinova and D.R. Flower, Towards the quantitative prediction of T-cell epitopes: CoMFA and CoMSIA studies of peptides with affinity to class I MHC molecule HLA-A*0201, *J. Med. Chem.*, 2001, **44**, 3572–3581.
126. I. Doytchinova and D.R. Flower, Physicochemical explanation of peptide binding to HLA-A*0201 Major Histocompatibility Complex. A three-dimensional quantitative structure–activity relationship study, *Proteins*, 2002, **48**, 505–518.
127. S.M. Free, Jr. and J.W. Wilson, A mathematical contribution to structure–activity studies, *J. Med. Chem.*, 1964, **7**, 395–399.
128. I.A. Doytchinova, M.J. Blythe and D.R. Flower, Additive method for the prediction of protein–peptide binding affinity. Application to the MHC Class I molecule HLA-A*0201, *J. Proteome Res.*, 2002, **1**, 263–272.
129. I.A. Doytchinova and D.R. Flower, Quantitative approaches to computational vaccinology, *Immun. Cell Biol.*, 2002, **80**, 270–279.
130. A. Kurata and J.A . Berzofsky, Analysis of peptide residues interacting with MHC molecule or T cell receptor. Can a peptide bind in more than one way to the same MHC molecule?, *J. Immunol.*, 1990, **144**, 4526–4535.
131. B. Gopalakrishnan and B.P. Roques, Do antigenic peptides have a unique sense of direction inside the MHC binding groove? A molecular modelling study, *FEBS Lett.*, 1992, **303**, 224–228.
132. C. Zhang, A. Anderson and C. DeLisi, Structural principles that govern the peptide-binding motifs of class I MHC molecules, *J. Mol. Biol.*, 1998, **281**, 929–947.
133. A.C. Young, S.G. Nathenson and J.C. Sacchettini, Structural studies of class I major histocompatibility complex proteins: insights into antigen presentation. *FASEB J.*, 1995, **9**, 26–36.
134. G. Chelvanayagam, I.B. Jakobsen, X. Gao and S. Easteal, Structural comparison of major histocompatibility complex class I molecules and homology modelling of five

distinct human leukocyte antigen-A alleles, *Protein Eng.*, 1996, **9**, 1151–1164.
135. G. Chelvanayagam, A roadmap for HLA-A, HLA-B, and HLA-C peptide binding specificities, *Immunogenetics*, 1996, **45**, 15–26.
136. G.A. Chelvanayagam, roadmap for HLA-DR peptide binding specificities, *Hum. Immunol.*, 1997, **58**, 61–69.
137. A. Baas, X. Gao and G. Chelvanayagam, Peptide binding motifs and specificities for HLA-DQ molecules, *Immunogenetics*, 1999, **50**, 8–15.
138. I.B. Jakobsen, X. Gao, S. Easteal and G. Chelvanayagam, Correlating sequence variation with HLA-A allelic families: implications for T cell receptor binding specificities, *Immunol. Cell Biol.*, 1998, **76**, 135–142.
139. O. Schueler-Furman, Y. Altuvia and H. Margalit, Examination of possible structural constraints of MHC-binding peptides by assessment of their native structure within their source proteins, *Proteins*, 2001, **45**, 47–54.
140. J. Aqvist, C. Medina and J.E. Samuelsson, A new method for predicting binding affinity in computer-aided drug design, *Protein Eng.*, 1994, **7**, 385–391.
141. R. Rosenfeld, Q. Zheng, S. Vajda and C. DeLisi, Flexible docking of peptides to class I major-histocompatibility-complex receptors, *Genet. Anal.*, 1995, **12**, 1–21.
142. U. Sezerman, S. Vajda and C. DeLisi, Free energy mapping of class I MHC molecules and structural determination of bound peptides, *Protein Sci.*, 1996, **5**, 1272–1281.
143. G. Vasmatzis, C. Zhang, J.L. Cornette and C. DeLisi, Computational determination of side chain specificity for pockets in class I MHC molecules, *Mol. Immunol.*, 1996, **33**, 1231–1239.
144. C. Zhang, J.L. Cornette and C. Delisi, Consistency in structural energetics of protein folding and peptide recognition, *Protein Sci.*,1997, **6**, 1057–1064.
145. Z. Weng and C. DeLisi, Toward a predictive understanding of molecular recognition, *Immunol. Rev.*, 1998, **163**, 251–266.
146. D. Rognan, M.J. Reddehase, U.H. Koszinowski and G. Folkers, Molecular modeling of an antigenic complex between a viral peptide and a class I major histocompatibility glycoprotein, *Proteins*, 1992, **13**, 70–85.
147. D. Rognan, N. Zimmermann, G. Jung and G. Folkers, Molecular dynamics study of a complex between the human histocompatibility antigen HLA-A2 and the IMP58–66 nonapeptide from influenza virus matrix protein, *Eur. J. Biochem.*, 1992, **208**, 101–113.
148. D. Rognan, L. Scapozza, G. Folkers and A. Daser, Molecular dynamics simulation of MHC-peptide complexes as a tool for predicting potential T cell epitopes, *Biochemistry*, 1994, **33**, 11476–11485.
149. P. Kern, R.M. Brunne, D. Rognan and G. Folkers, A pseudo-particle approach for studying protein-ligand models truncated to their active sites. *Biopolymers*, 1996, **38**, 619–637.
150. D. Rognan, S. Krebs, O. Kuonen, J.R. Lamas, J.A. Lopez de Castro and G. Folkers, Fine specificity of antigen binding to two class I major histocompatibility proteins (B*2705 and B*2703) differing in a single amino acid residue, *J. Comput. Aided Mol. Des.*, 1997, **11**, 463–478.
151. S. Krebs, D. Rognan and J.A. Lopez de Castro, Long-range effects in protein-ligand interactions mediate peptide specificity in the human major histocompatibilty antigen HLA-B27 (B*2701), *Protein Sci.*, 1999, **8**, 1393–1399.
152. D. Rognan, L. Scapozza, G. Folkers and A. Daser, Rational design of nonnatural peptides as high-affinity ligands for the HLA-B*2705 human leukocyte antigen, *Proc. Nat. Acad. Sci. USA*, 1995, **92**, 753–757.

153. S. Krebs and D. Rognan, From peptides to peptidomimetics: design of nonpeptide ligands for major histocompatibility proteins, *Pharm. Acta Helv.*, 1998, **73**, 173–181.

154. S. Dedier, S. Krebs, J.R. Lamas, S. Poenaru, G. Folkers, J.A. Lopez de Castro, D. Seebach and D. Rognan, Structure-based design of nonnatural ligands for the HLA-B27 protein, *J. Recept. Signal Transduct. Res.*, 1999, **19**, 645–657.

155. A. Caflisch, P. Niederer and M. Anliker, Monte Carlo docking of oligopeptides to proteins, *Proteins*, 1992, **13**, 223–230.

156. J.S. Lim, S. Kim, H.G. Lee, K.Y. Lee, T.J. Kwon and K. Kim, Selection of peptides that bind to the HLA-A2.1 molecule by molecular modelling, *Mol. Immunol.*, 1996, **33**, 221–230.

157. I.P. Androulakis, N.N. Nayak, M.G. Ierapetritou, D.S. Monos and C.A. Floudas, A predictive method for the evaluation of peptide binding in pocket 1 of HLA-DRB1 via global minimization of energy interactions, *Proteins*, 1997, **29**, 87–102.

158. H. Toh, N. Kamikawaji, T. Tana, S. Muta, T. Sasazuki and S. Kuhara, Magnitude of structural changes of the T-cell receptor binding regions determine the strength of T-cell antagonism: molecular dynamics simulations of HLA-DR4 (DRB1*0405) complexed with analogue peptide, *Protein Eng.*, 2000, **13**, 423–429.

159. N. Froloff, A. Windemuth and B. Honig, On the calculation of binding free energies using continuum methods: application to MHC class I protein-peptide interactions, *Protein Sci.*, 1997, **6**, 1293–1301.

160. N. Arora and D. Bashford, Solvation energy density occlusion approximation for evaluation of desolvation penalties in biomolecular interactions, *Proteins*, 2001, **43**, 12–27.

161. O. Schueler-Furman, R. Elber and H. Margalit, Knowledge-based structure prediction of MHC class I bound peptides: a study of 23 complexes, *Fold Des.*, 1998, **3**, 549–564.

162. M. De Maeyer, J. Desmet and I. Lasters, The dead-end elimination theorem: mathematical aspects, implementation, optimizations, evaluation, and performance, *Methods Mol. Biol.*, 2000, **143**, 265–304.

163. D. Rognan, S.L. Lauemoller, A. Holm, S. Buus and V. Tschinke, Predicting binding affinities of protein ligands from three-dimensional models: application to peptide binding to class I major histocompatibility proteins, *J. Med. Chem.*, 1999, **42**, 4650–4658.

164. A. Logean, A. Sette and D. Rognan, Customized versus universal scoring functions: application to class I MHC-peptide binding free energy predictions, *Bioorg. Med. Chem. Lett.*, 2001, **11**, 675–679.

165. P. Kangueane, M.K. Sakharkar, K.S. Lim, H. Hao, K. Lin, R.E. Chee and P.R. Kolatkar, Knowledge-based grouping of modeled HLA peptide complexes, *Hum. Immunol.*, 2000, **61**, 460–466.

166. Y. Altuvia, O. Schueler and H. Margalit, Ranking potential binding peptides to MHC molecules by a computational threading approach, *J. Mol. Biol.*, 1995, **249**, 244–250.

167. Y. Altuvia, A. Sette, J. Sidney, S. Southwood and H. Margalit, A structure-based algorithm to predict potential binding peptides to MHC molecules with hydrophobic binding pockets, *Hum. Immunol.*, 1997, **58**, 1–11.

168. S. Miyazawa and R.L. Jernigan, Residue–residue potentials with a favorable contact pair term and an unfavorable high packing density term, for simulation and threading, *J. Mol. Biol.*, 1996, **256**, 623–644.

169. O. Schueler-Furman, Y. Altuvia, A. Sette and H. Margalit, Structure-based prediction of binding peptides to MHC class I molecules: application to a broad range of

MHC alleles, *Protein Sci.*, 2000, **9**, 1838–1846.

170. M.R. Betancourt and D. Thirumalai, Pair potentials for protein folding: choice of reference states and sensitivity of predicted native states to variations in the interaction schemes, *Protein Sci.*, 1999, **8**, 361–369.

171. M.T. Swain, A.J. Brooks and G.J.L. Kemp, An automated approach to modelling class II MHC alleles and predicting peptide binding, *Proceedings of the Second IEEE International Symposium on Bio-Informatics and Biomedical Engineering*, IEEE Computer Society Press (in press).

172. S. Lu, V.E. Reyes, C.M. Bositis, T.G. Goldschmidt, V. Lam, R.R. Torgerson, T. Ciardelli, L. Hardy, R.A. Lew and R.E. Humphreys, Biophysical mechanism of the scavenger site near T cell-presented epitopes, *Vaccine*, 1992, **10**, 3–7.

173. C. Raychaudhury, A. Banerjee, P. Bag and S. Roy, Topological shape and size of peptides: identification of potential allele specific helper T cell antigenic sites, *J. Chem. Inf. Comput. Sci.*, 1999, **39**, 248–254.

174. M.P. Davenport, I.A. Ho Shon and A.V. Hill, An empirical method for the prediction of T-cell epitopes, *Immunogenetics*, 1995, **42**, 392–397.

175. U. Hobohm and A. Meyerhans, A pattern search method for putative anchor residues in T cell epitopes, *Eur. J. Immunol.*, 1993, **23**, 1271–1276.

176. Y. Altuvia, J.A. Berzofsky, R. Rosenfeld and H. Margalit, Sequence features that correlate with MHC restriction, *Mol. Immunol.*, 1994, **31**, 1–19.

177. A.M. Livingstone and C.G. Fathman, The structure of T-cell epitopes, *Annu. Rev. Immunol.*, 1987, **5**, 477–501.

178. A.J. Deavin, T.R. Auton and P.J. Greaney, Statistical comparison of established T-cell epitope predictors against a large database of human and murine antigens, *Mol. Immunol.*, 1996, **33**, 145–155.

179. C.J. Stille, L.J. Thomas, V.E. Reyes and R.E. Humphreys, Hydrophobic strip-of-helix algorithm for selection of T cell-presented peptides, *Mol. Immunol.*, 1987, **24**, 1021–1027.

180. J.B. Rothbard and W.R. Taylor, A sequence pattern common to T cell epitopes, *EMBO J.*, 1988, **7**, 93–100.

181. J.L. Cornette, H. Margalit, J.A. Berzofsky and C. DeLisi, Periodic variation in side-chain polarities of T-cell antigenic peptides correlates with their structure and activity, *Proc. Natl. Acad. Sci. USA*, 1995, **92**, 8368–8372.

182. M.H. Van Regenmortel and G. Daney de Marcillac, An assessment of prediction methods for locating continuous epitopes in proteins, *Immunol. Lett.*, 1988, **17**, 95–107.

183. Md. Ferreira-da-Cruz, S. Giovanni-de-Simone, D.M. Banic, M. Canto-Cavalheiro, D. Camus and C.T. Daniel-Ribeiro, Can software be used to predict antigenic regions in Plasmodium falciparum peptides?, *Parasite Immunol.*, 1996, **18**, 159–161.

184. J.M. Thornton, M.S. Edwards, W.R. Taylor and D.J. Barlow, Location of 'continuous' antigenic determinants in the protruding regions of proteins, *EMBO J.*, 1986, **5**, 409–413.

185. D.J. Barlow, M.S. Edwards and J.M. Thornton, Continuous and discontinuous protein antigenic determinants, *Nature*, 1986, **322**, 747–748.

186. M.H. Van Regenmortel and J.L. Pellequer, Predicting antigenic determinants in proteins: looking for unidimensional solutions to a three-dimensional problem?, *Pept. Res.*, 1994, **7**, 224–228.

187. A.J. Alix, Predictive estimation of protein linear epitopes by using the program PEOPLE, *Vaccine*, 1999, **18**, 311–314.

188. J.L. Pellequer and E. Westhof, PREDITOP: a program for antigenicity prediction,

J. Mol. Graph., 1993, **11**, 204–210.

189. M. Gromiha, A simple method for predicting transmembrane helices with better accuracy, *Prot. Eng.*, 1999, **12**, 557–561.

190. K. Nishikawa and R. Apweiler, *J. Biochem.*, 1982, **91**, 1821–1824.

191. F. Eisenhaber, C. Frömmel and P. Argos, Prediction of secondary structural content of proteins from their amino acid composition alone. II. The paradox with secondary structural class, *Proteins*, 1996, **25**, 169–179.

192. H. Chiapello *et al.*, Codon usage as a tool to predict the cellular location of eukaryotic ribosomal proteins and aminoacyl-tRNA synthetases, *Nucleic Acids Res.*, 1999, **27**, 2848–2851.

193. K. Chou and D. Elrod, Using discriminant functions for prediction of subcellular location of prokaryotic proteins, *Biochem. Biophy. Res. Comm.*, 1998, **252**, 63–68.

194. J. Cedano, P. Aloy, J.A. Perez-Pons and E. Querol, Relation between amino acid composition and cellular location of proteins, *J. Mol. Biol.*, 1997, **266**, 594–600.

195. H. Nakashima, Discrimination of intracellular and extracellular proteins using amino acid composition and residue-pair frequencies, *J. Mol. Biol.*, 1994, **238**, 54–61.

196. M. Andrade, S. O'Donoghue and B. Rost, Adaptation of protein surface to subcellular location, *J. Mol. Biol.*, 1998, **276**, 517–525.

197. A. Reinhardt and T. Hubbard, Using neural networks for prediction of the subcellular location of proteins, *Nucleic Acid Res.*, 1998, **26**, 2230–2236.

198. K. Nakai, Predicting various targeting signals in amino acid sequences, *Chem. Res. Kyoto Univ.*, 1991, **69**, 269–291.

199. K. Nakai, Protein sorting signals and prediction of subcellular location, *Adv. Protein Chem.*, 2000, **54**, 277–345.

200. M.S. Briggs, L.M. Gierasch, A. Zlotnick, J.D. Lear and W.F. DeGrado, In vivo function and membrane-binding properties are correlated for Escherichia coli LamB signal peptides, *Science*, 1985, **228**, 1069–1099.

201. G. von Heijne, Signal sequences, the limits of variation, *J. Mol. Biol.*, 1985, **184**, 99–105.

202. B. Margolti and B. Dobberstein, Signal sequences, more than just greedy peptides, *Trends Cell Biol.*, 1998, **8**, 410–415.

203. M. Sjöström, S. Wold, A. Wiesland and L. Rilfos, Signal peptide amino acid sequences in Escherichia coli contain information related to final protein localisation: a multivariate data analysis, *EMBO J.*, 1987, **6**, 823–831.

204. M. Edman, T. Jarhede, M. Sjöström and A. Wiesland, Different sequence patterns in signal peptides from mycoplasms, other Gram positive bacteria and Escherichia: a multivariate data analysis, *Proteins: Struct., Funct., Genet.*, 1999, **35**, 195–205.

205. H. Nielsen, J. Engebrecht, S. Brunak and G. von Heijne, Identification of eukaryotic signal peptides and prediction of their clevage sites, *Protein Eng.*, 1995, **1**, 1–6.

206. B. Jagla and J. Schuchhardt, Adaptive encoding neural networks for the recognition of human signal peptide cleavage sites, *Bioinformatics*, 2000, **16**, 245–250.

207. J. Lu and E. Celis, Use of two predictive algorithms of the world wide web for the identification of tumor-reactive T-cell epitopes, *Cancer Res.*, 2000, **60**, 5223–5227.

208. M.H. Andersen, L. Tan, I. Sondergaard, J. Zeuthen, T. Elliott and J.S. Haurum, Poor correspondence between predicted and experimental binding of peptides to class I MHC molecules, *Tissue Antigens*, 2000, **55**, 519–531.

209. J.M. Pezzuto, Plant-derived anticancer agents, *Biochem. Pharmacol.*, 1997, **53**, 121–133.

Subject Index